Lewis Anderson

Natural way in diet

The proper food of man

Lewis Anderson

Natural way in diet
The proper food of man

ISBN/EAN: 9783337201340

Printed in Europe, USA, Canada, Australia, Japan

Cover: Foto ©berggeist007 / pixelio.de

More available books at **www.hansebooks.com**

"Nature" Series, No. 1.

NATURAL WAY IN DIET

OR

THE PROPER FOOD OF MAN

BY

PROF. L. H. ANDERSON

Author of "How to Win," "Occult Forces," "Key to Power," Etc., Etc.

"Prove all things, and hold fast that which is good."

PUBLISHED BY

THE NATIONAL INSTITUTE OF SCIENCE,

MASONIC TEMPLE, CHICAGO, ILL.

1898.

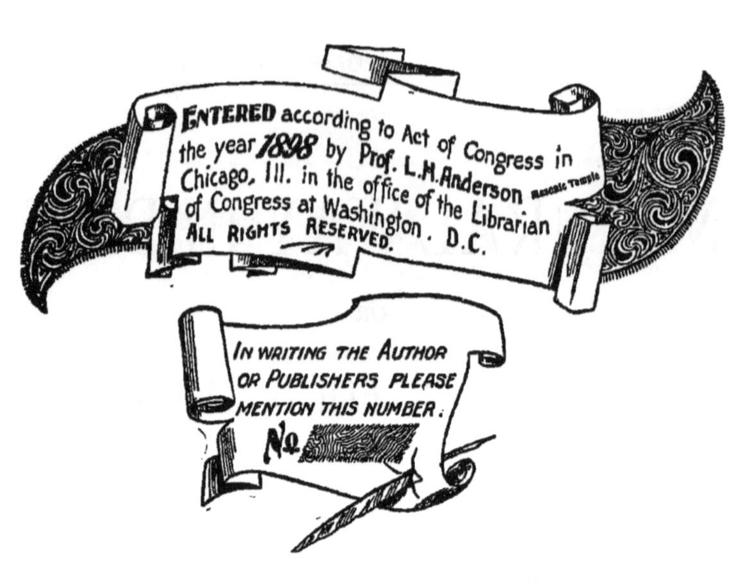
ENTERED according to Act of Congress in
the year 1898 by Prof. L.H. Anderson Masonic Temple
Chicago, Ill. in the office of the Librarian
of Congress at Washington. D.C.
ALL RIGHTS RESERVED.
IN WRITING THE AUTHOR
OR PUBLISHERS PLEASE
MENTION THIS NUMBER:
№

CONTENTS.

CHAPTER V.

Cruelty, Depraved Appetites, Disease and Death.

CHAPTER VI.

Vivisection.

CHAPTER VII.

STUDIES IN FOODS.

CHAPTER VIII.

EFFECTS OF DIFFERENT FOODS.

CHAPTER IX.

Opinions of the Ancients.

CHAPTER X.

Opinions of Modern Thinkers.

INTRODUCTION.

During years of work and study as a practical healer, I have been lead to realize the importance of a book which would teach the people the value of foods, and the intimate relation existing between the food taken into the system and the higher life of man. The subject is one which should be of deep interest to every student of human life, and to every one who is seeking to come into harmonious relation with the laws of Nature. It rests upon a scientific basis, and the workings of the laws of cause and effect are as plainly discernible as in any other process of Nature.

It has been my earnest endeavor to lead all those with whom it has been my privilege to come in contact, as teacher, physician and friend, to new and better methods of thinking; to a higher, truer, holier life. As Principal of the National Institute of Science, my field of labor has been wide, and the loving testimonies of students, whose lives have been broadened and sweetened by my efforts in their behalf, is sufficient reward for my earnest labors, and ample evidence of my success.

In the course of lessons in Personal Magnetism, the student is taught how to care for his body, that temple of the living soul, and is shown its true relation to the psychic or higher life, of which it is but the instrument. It should be kept in perfect condition if it is expected to perform its proper functions. The man who neglects his body, neglects at the same time his higher spiritual nature. The man who abuses his body in any way violates his inmost soul.

It can not be denied by any one who has given the slightest thought to the subject of the influence of foods upon the functions of body and mind, that a pure diet will result in a pure body, a pure body in pure thoughts, higher ideals, and nobler aspirations. If the system is furnished with healthful material with which to build, and the delicate organs perform their work in a natural manner, the whole body will be clean in every department. If every law of Nature is obeyed, the whole human organism will be perfect in every respect. It is by disobedience that man brings disease, weakness and sin into his life. Every pang suffered, whether of physical or mental anguish, is the direct result of a broken law. For every broken law there is its attendant and consequent evil affecting the body corporal and the body spiritual.

As a result of gross and unnatural habits of life we have disease, poverty, sin. As a result of these conditions we have an army of medical and spiritual advisers, who must be kept in luxury, and whose purses must be filled at the expense of human happiness. We have prisons and penitentiaries, houses of correction and hospitals, poor houses and orphan asylums. We have misery, degradation and vice everywhere. Go where you will it confronts you. Every cause is followed by an effect, every effect is in itself a cause to produce still other effects. Every effect is good in proportion as the cause which produces it is good.

It has been my highest aim to teach people how to set in motion those causes which will bring into their lives harmony and happiness, health, strength and contentment. Realizing that improper foods create abnormal appetites, I have endeavored to point them to a perfect way in diet, which would bring man into a state of purity, and make his life what it should be. I have studied carefully the effects of foods and drinks upon individual and national life. The results of my observations, extending over a long period of time, have been set forth in this book.

My object has been to help the human race to realize the grossness of its present life, to teach them to despise that which is degrading in its effects, and to follow that which is good and true. I feel sure that thousands, after having been pointed to this better way of living, will feel the importance of entering upon the path which leads to a nobler state. I have been actuated by a sincere desire to do good, and feel assured that my efforts will not have been in vain.

So long as human wrecks float upon the vast sea of life, so long as man's hopes are blasted by ignorance, so long as human hearts are wrung with the agony of despair and grim poverty grins at the doorway of the unfortunate, and vice slinks in the shadows of superstition, so long will I, by every effort of voice and pen, preach the gospel of truth, temperance and cleanliness; so long will I proclaim the glad tidings of higher, better things for the earnest seeker of happiness, contentment and universal health.

So long as man is in ignorance of the subtle laws that control him, he must needs look to others for truth; but when he studies self and can trace the invisible working of the forces which produce the conditions surrounding him, he will have no further need of the

Master, but will be able to control wisely and well his destiny. I trust the dawn is near at hand.

The beginner in the new life will need much practical, personal advice, such as will be given to our students in Personal Magnetism. This course of study is designed with the purpose of developing in the most complete and harmonious manner, the human body and the mental powers. It is praised and advised by men of science, whose wide knowledge and clear judgment makes them authority upon questions relating to the science of human life.

In reference to the diet we advocate in this book, we defy any one to offer any objection to it which can not be successfully combated by experience, reason, science and truth. We have often been personally assailed by those whose ideas did not fully coincide with our own, and who thought that by argument they would be able to tear down the theories advanced by the most learned and scientific men of every age. But truth, no matter where found, will withstand the gibes of the persecutor. Medical men can no longer stumble blindly in the pathway that has been marked out for them, for the light of the Sun of Science casts its rays athwart the way, and forces them to see the truth. They begin to realize that there is a wisdom not of men which will teach them to listen to Nature, and to teach others to seek of her the blessings they so much crave.

Men to-day seek for themselves in the temple of truth the answer from the oracles of their goddess—Nature. They no longer ask the physician to interpret for them, but each harkens to her voice and reads for himself the mystic interpretation of health.

Those who feel it impossible to take our course in Personal Magnetism, and who by reading these pages may determine to give up a flesh diet, have taken a step in the right direction, and I know that after giving it a thorough trial, they will never go back to the old diet, with its gross impurities, which poison the whole system and make life a burden. He who abstains from meats does well, but he who abstains entirely from the sensualities of the table, and lives only on the pure diet of Nature, does better. There are many who would never again taste flesh did they but fully realize the fearful effects it produces upon the system, and who would find nothing objectionable in a purely vegetable diet for a long time—not until the grossness of his physical and psychic nature had been subdued; but there is a still higher plane, and until one has overcome the inherited and acquired tendencies of generations of sensuality he will be unable to appreciate it. Our students are lead up step by step from the lower to the higher, and in time their forces are so developed, their faculties so sensitive to the higher harmonies of life, that the things once loved are now esteemed but lightly.

He who attains to this state is indeed fortunate. He who by his life influences others to live a holier life has done a noble deed, though it was done unconsciously. The subject of diet is broad, too broad to handle in a book of this size. As gluttony produces drunkenness and disease, sin and suffering, so temperance in eating and drinking produces a corresponding purity of life and thought. As the gross man influences the lives of those about him, so will the moral man influence those who come within his radius, and help to mould their lives for higher things than mere sensualities.

This thought should be an inspiration to all to live purely, for not only your own welfare depends upon your method of life, but the welfare of those you meet in daily life. So long as you travel in the old way of ignorance, you are keeping others in the same path; but when you step out into the sunlight of freedom, you embolden others to follow. There are many who, while not daring to think for themselves along new lines, are yet willing to accept the doctrines after they have once been proclaimed by a leader.

You should never be content to be a mere follower. Be independent enough to think for yourself, and when you have once become convinced that any theory is wrong, no matter how plausible the argument in favor of it, do not be afraid to reject it, and trust to your own good sense and calm reason rather than to the opinions of others. Wherever common sense leads you need not fear to follow. Study your subject and thoroughly understand it, then do not allow any one else to decide what you shall do. Custom and fashion are arbitrary, but you can be superior to both. After you have given the subject of foods due study, and have seen the terrible results upon the human race of flesh-eating, your decision will be in favor of a natural diet, and after you have lived awhile upon the pure products, you will have no desire for those things which you know debase your whole nature.

I have never yet met one who, after abstaining from meat for a period, was willing to return to the gross flesh and grease of former days, the heavy, sodden vegetables and indigestible bread that made his life a burden and deprived him of the joys of life. He knows that his system is now clean, and that when he eats those things which Nature has given, that he is not taking into his system some loathsome disease. He knows that he is taking into his system the elements best adapted to him, the material which will produce pure blood, pure thoughts, and chaste desires. He knows his life will be sweeter, truer and nobler from his temperance in eating, for this appetite controls all others.

I trust that he who reads this book may be led to give serious thought to the subject of foods and their relation to human life. That he will observe the truth of the statements here given in the lives of those about him, and that he may be lead to abandon forever the gross and filthy habits of eating and drinking, and substitute for the hot and heavy pastries, greasy vegetables and half-cooked carcasses of murdered animals the delicacies so bountifully provided by a lavish Mother for his sustenance.

The years I have devoted to the study of human life have enabled me to speak with no uncertain voice against the prevailing vice of the people. I have given my life to the human race without reserve, and have spared myself no trouble, no expense, no amount of personal toil to teach them the best methods of living this life, believing that in so doing I was accomplishing the highest mission, for when I have studied the ills which make life miserable to so many thousands of my fellow-men, I have found the foundation of them all were the gross habits in which they indulged; therefore, I have always preached to them the gospel of temperance and cleanliness. My creed has been short, but no one can fail to understand its tenets. My earnest endeavor has been to bring man into harmony with Nature, to teach him to obey her divine laws. Knowing that sickness, disease, poverty, misery and crime were unnatural conditions, and knowing, too, that all unnatural conditions are unnecessary, I have ever labored to establish those conditions which would lift him out of his present groveling, sensual life and place him upon that plane which is the common heritage of all.

With this purpose in view I have published "The Nature Series," which are designed to educate and develop man's whole being, to perfect his health and teach him to live in divine accord with those laws which if obeyed will make his life harmonious and beautiful. A well-rounded character is to be desired by all, and no one can attain it in its perfection unless he is physically well and has developed those interior forces which give him power over self and over circumstances.

Trusting that what I may say will cause you to "prove all things, and hold fast that which is good," I am, Very sincerely,

L. H. ANDERSON.

CHAPTER I.

MAN NOT NATURALLY A CARNIVORA.

THE welfare of a nation depends very largely upon its diet and knowledge of hygiene, and it is a serious mistake for a highly civilized State to allow its masses to grow up in ignorance of the fundamental laws of health, of what constitutes wholesome, economical and nutritious food, and of the proper methods of preparing the same.

"Tell me what you eat, and I will tell you what you are," say some of the most radical dietic reformers of the day. This is partly true. If a man is a glutton, devouring quantities of coarse food, and washing it down with abominable liquids, his every look and action will testify to the grossness of his physical fiber. His thoughts and appetite will be no less coarse than the food he eats. His life, which is but the outward expression of the inward thought, must of necessity become brutalized and degraded by the common material of which it is built.

As a nation we have paid no attention whatever to diet. The rich have surfeited on the most exquisite (?) viands procurable. The feasts of Lucullus have faded into insignificance before the royal magnificence of our gluttonous riots. For our delectation men have searched sea and shore, braved the terrors of the northern blast and toiled unceasingly under torrid skies; they have slaughtered the innocent and robbed the helpless, destroyed with wanton savagery and laid the trophies at our feet to satisfy our abnormal appetites. The purple grape has yielded its rich blood to please our taste, and we have manufactured hideous poison from every grain and fruit to burn with unquenchable fire the delicate tissues of the body. The helpless victims of our insatiable lust are countless. In every part of the land we have erected immense shambles, where daily sacrifice is made to the Moluch of our beastly passion. We have built high wine vats, where is poured out constant oblations to our gods of strong drink. We have, in our ignorance, made us a nation of degraded brutes. Swayéd only by impulse, victims of uncontroled appetites, can we wonder that man, created in the image of the Divine, has fallen to the level of the brute creation?

Man, the master of the world, has made it what it is. Into his hands it was delivered with all its mighty possibilities for good, for evil, for happiness and for woe. Yielding to the admonitions of his sensual nature, he has made of it a very hell, which consumes the inmost soul of the tender and the loving with anguish unquenchable. Would he but work in harmony with the divine laws of Nature, he could convert it into the original Paradise of man.

When we would study the methods and laws which philosophy and modern science have indicated as best adapted to the development and perfection of our kind, we turn first to natural history, and seek in the study of comparative anatomy of men and other animals for information regarding the habits and mode of living of primitive man.

The classification of Linnæus, which is admitted without serious objections by eminent scientific men to be the correct one, we find under the name of Primates the highest order in the class of mammiferous animals, and at the head of this classification is placed man and the anthropoid ape. The last contains two species, which, considered from an anatomical and physiological point, very closely resembles the human family.

In the endeavor to decide what ought to be, consistently with the wise laws of nature, the habits and diets of man, we naturally inquire what mode of alimentation is best suited to the animals most nearly resembling him, and proceed to analyze first their physical structure.

A most superficial observation enables us to recognize the resemblance existing between the general conformation of the skull of the ape and that of man. The nervous system, which dominates the functions of all the organs, repairing lesions, regulating the processes of every system, presiding over the harmony of their operations, preserver and law-giver of the bodily kingdom, is a prime importance, and above all the dominant portion of the animal system. Those animals in which the nervous system most closely resembles the human type in its development, will possess the right to be classed nearest to man in the animal kingdom. Upon closer study we find the orang-outang most closely imitates him, from the fact that the perfection of the nervous system—and in particular the development of its ganglionic centers—is second in degree to man's, in whom we find the supreme aggregation and complete composition of the parts which constitute this system.

Eminent scientists have concluded that the sole difference between the orang-outang and the human subject is one not of degree but of kind, and it is a well-known fact that the dental morphology and formula of the apes of the old world are identical with those of man, while the same formation is met with in the surface of the molar teeth, and in the superficial disposition of the enamel in man, as is found in the orang-outang, the chimpanzee and the gorilla. Thus it will be seen that a strong resemblance exists between the structure of man and the lower animals, while the many smaller points of resemblance place them in close relation to the human family.

At first sight it is barely possible to distinguish between the stomach of man and the stomach of the higher ape. The human stomach is somewhat smaller, however; the entire digestive organism of the ape is very similar to that of man, while the stomach of the carnivorous quadrupeds differ very materially from that of the human family, not only in regard to its relative dimensions, but also as to its form.

We do not propose to enter into wearisome detail as to the difference between the structure of man as compared with that of other animals. We wish merely to combat certain erroneous impressions which have obtained credence, not only among the unlearned, but also among those who are supposed to be authority upon scientific subjects. For example, have we not often heard people speak with all the authority of conviction of the "canine teeth" and "simple stomach" of man as being unquestionable evidence of his natural adaptation for a flesh diet? We have wished to demonstrate but one fact: If the arguments of our so-called authorities are valid, they must apply with even greater force to the anthropoid apes, whose "canine" teeth are much longer and more powerful than those of man.

Man is neither carnivorous nor herbivorous. His teeth, as well as his digestive organism, are entirely different from those animals which subsist upon herbs and flesh. We must, therefore, conclude that he is by nature and origin frugivorous, as is the ape, which he so nearly resembles.

While the objection may possibly be offered that since man is, according to natural structure and propensities, a fruit and seed-eater, he should confine himself to these rather than partake so freely of the leguminous plants and roots which belong to the dietary of the herb-eaters, whose organization differs so greatly from that of man; yet we cannot but believe that a combination of natural and artificial forces has made it impossible for the majority of people of many parts of the globe to subsist without the art of cooking. It seems, therefore, both wise and consistent that they should increase the variety and range of their food by cooking. We, however, believe that fire can be used legitimately by man only for the preparation of food which otherwise he would not be able to masticate.

The true frugivora does not refuse to partake of a diet composed of cooked fruits, vegetables and roots, but rather enjoys them. In their uncooked state they are not distasteful to the sight, smell nor palate. With man Art has superseded Nature to such an extent that he is enabled, by means of baking, boiling and stewing, and by use of a multiplicity of disguises, to eat without disgust and digest without discomfort the food of the tiger, bear, wolf and hyena. This abuse of the art of cookery has degraded man to the level of the beasts of prey, and has made him a savage monster, feasting upon his fellow-creatures.

The preparation of the soil, the culture of plants, harvesting and garnering, and the use of vegetable produce, are alike in harmony with the interests of the highest morality of individual and national health, of economy and of love, of truth, beauty and purity.

We have shown that mankind are by nature frugivorous, and it has been demonstrated that they can also become omnivorous and carnivorous. Let us now inquire, therefore, whether such transformation of his nature has been attended with any material advantage to the individual or to the race as a whole.

The belief that flesh-food contains all the elements of physical force, and that in order to be strong physically and bright mentally man must partake very largely of animal food is as common as it is false. While it has obtained credence not only among the vulgar and unscientific, physicians and teachers everywhere have fallen into the same error, not as a result of their scientific discoveries, but because of accepted authority and the gross customs of the people. We, nevertheless, see these opinions daily disproven, for the strongest and most useful animals are those which live entirely upon a non-meat diet. They far surpass in force and endurance their flesh-eating masters. The faithful horse, the patient ox, the mule, the camel and the elephant, beasts which are in daily use, and without which it would be impossible for us to accomplish the vast enterprises incident to civilized life, all derive their magnificent strength from the fruits of the soil. The buffalo, the bison and the hippopotamus are splendid types of physical power and endurance, while the gorilla, nourished by fruits and nuts almost exclusively, is dreaded, feared by the explorer and naturalist on account of its immense strength.

Those nations which have left to us the most glorious

records of their greatness, the most superb monuments of their prowess, the profoundest and purest thought, the sweetest songs and the highest state of civilization, were not a nation of meat-eaters.

In the palmiest days of the Greek and Roman empires, before their people were debauched by intemperance and licentiousness, their soldiers and heroes subsisted on the simplest vegetable food. The chief food of the gladiators, who have long been renowned in song and story, was barley-cakes and oil. How would our modern athletes like this diet upon which the wrestlers and runners of antiquity grew strong and perfect physically, and gained self-control and mental power?

The greater part of the population of China and Japan are devoted to the teachings of Buddah, and they not only condemn the use of animal flesh as an article of diet, but show peculiar kindness and tenderness to the brute creation. The murderer of animals is severely punished. We can imagine the horror of one of these men, taught to consider all life sacred, when suddenly set down in the midst of our boasted civilization to behold upon every hand evidence of our savagery and brutality. He hears the piteous cries of the innocent victim as it is driven to the sacrifice; he gazes at the huge altars of lust with amazement depicted in every line of his face; he beholds with undisguised horror the slaughter of the helpless, and his heart is filled with loathing for a people who thus devour their fellow-creatures. He sees no beauty in our "Holy days," where devotees first worship a god, and then slay the work of his hands. Our feasts are an abomination to him, with their untold multitudes of slaughtered innocents. They turn with disgust from our table, groaning under a load of cooked carcasses, surrounded by gravies; rich compounds of pastries, tankards of ale, beer and wines, to their own frugal fare of brown bread and fruits, melons and nuts.

By examination it is found that the peasantry, which represents the bone and brawn of every nation, subsists almost entirely upon vegetable diet, and are stronger, more enduring, and, where they have an opportunity of proving their real worth, more clever mentally than are the higher classes who do not practice frugality.

Since, then, we have found that man by his external structure, his internal organism, his natural instincts and habits is a frugivorous animal, can we longer doubt that the fruits and herbs which Mother Nature so bountifully lavishes upon him is his proper food?

All the various alimentary substances are divided and subdivided into various substances. The various salts necessary for perfect alimentation are present in proper quantities. Potash, chlorides and phosphates are found in the cereals, and as they exist principally in the exterior portions of various grains, it is obvious that they are contained only in small quantities in the ordinary white breads. It is only when the flour is used unbolted that we obtain the full nutritive value of the grain of which it is composed. From water the system derives various salts and minerals necessary to its perfect nutrition. Fresh fruits, apples, currents, citron, cranberries, cherries, grapes, plums, peaches and lemons contain large proportions of acids, salts, iron and water.

Thus we see that vegetable products contain all the elements necessary to the perfect nutrition of the body, and that in greater proportion than can be found in meat diet. We find, too, that the value of vegetable

With Kindest Wishes
Yours for Truth and Right.
Jno. K. Anderson

food taken as a whole is much greater than that of flesh food.

The proportional quantity of each principle which should enter into the daily alimentation of man varies according to sex, circumstances and personal habits, the amount required being much greater during active and prolonged labor than during periods of repose or with moderate exercise. This fact is self-evident, and it has been proven by science.

It has also been demonstrated that the alimentation afforded by a fruit and vegetable diet is richer in nutritive properties, purer in quality, and, therefore, better fitted to the requirements of man than is the flesh of animals. This superiority of a vegetable diet over meat is adduced from the fact that those who follow it are not only stronger and healthier, but they have little or no tendency to indulge in alcoholic stimulants, tobacco, etc. Flesh produces an effect upon the system similar to that produced by alcohol. The waste of the body is increased. The powers of resistance are diminished, and the tendency to disease increased. The baneful effects upon the whole system can not be overestimated. To many it may seem an exaggeration to say that the use of flesh induces premature death, but those who have given the subject even superficial thought must admit that it hastens the arrival of old age, besides engendering unchastity and drunkenness, encouraging vice and excesses of all kinds. Physicians have noted that those tribes which have been used to a vegetable diet, after a feast composed of meats exhibit all the effects of having partaken of intoxicants. The essence of beef will also produce upon sensitive persons the same effect as liquors.

It is an established fact that wild animals which have been captured, if fed upon vegetables, grow less ferocious and quarrelsome, while if their diet be changed but a few days to meat in large quantities they show all their old traits of character. Domestic animals, too, are affected in the same way by a change in diet. Swine and dogs if fed entirely upon flesh grow quite dangerous.

We have already alluded to the deplorable effects of flesh-eating, and have pointed out alcoholism as perhaps the commonest and most pernicious. Few, even among medical men, realize the terrible effects of this habit (for habit it is) upon mankind. Many who do use their knowledge to their own advantage by prescribing meats and liquors to their patients, thus making them more susceptible to disease and producing abnormal conditions of the patient, compelling them to remain under medical treatment for diseases he himself has thus created. Temperance reformers, after careful study of the subject, aver that the use of flesh foods is the direct or indirect cause of much of the intemperance with which our land is cursed. The use of flesh excites the baser passions in man, keys the nervous system to a high tension and creates diseases as numerous as they are dangerous. The more flesh consumed the stronger the tendency for strong drink, and the greater the danger of confirmed alcoholism.

Physicians wisely act upon this fact in the treatment of dipsomania, the efficiency of their methods being due more to the strict regimen than to the curative properties of the remedies prescribed. Confirmed drunkards, who, becoming frightened at the future looming dark before them, enter a sanitarium for treatment, and at once experiencing satisfactory results, ascribe their cure to the magic pills or powders administered by the physician, and give no credit to the vapor-baths which eliminate the poison from the system ; the quiet, orderly, pleasant life which restores the nerves to their normal state ; the exercise required, which fills the veins with fresh blood, and enkindles anew the fires of life ; the regular diet, consisting of fruits and vegetables, which give the system pure material with which to build.

It is impossible for a man to live for a year or for even a few months upon a diet composed exclusively of ripe and lucious fruits, properly prepared vegetables and whole-wheat bread without eliminating from the entire system all craving for alcoholic stimulants.

Perhaps the most potent cause of prostitution is the indulgence of a perverted appetite for flesh and alcoholic stimulants. So long as people remain ignorant of the causes which produce the evils they are laboring to mitigate, so long must they needs work blindly, and so long will their efforts be crowned with failure. When the slaughter-pen, with its frightened, lowing victims, its horrors of blood and butchery, its swarms of employees, hardened to the suffering of innocence, is abolished, the distillery and the rum-shop, the house of prostitution, and the haunts of vice will disappear as by magic.

Then, too, the slaughter-pens, with their offensive odors, are recognized sources of disease, and yet they are held under but gentle surveillance. Our public authorities seemingly prefer to subject us constantly to the dangers arising from these pest spots than to do aught to remove from our midst the necessary (?) evils. It is entirely needless and morally unlawful that people of cultivation and refinement should have imposed upon them an occupation which is at once disgusting, brutal-

sumed meat for their food were forced to become their own butchers and prepare their victims for the table with their own hands, few would care to eat the flesh when it came to the table, while no one however delicately bred would object to plucking from vine and tree the bounties which kindly Nature provides.

The torture of the animals doomed to be sacrificed is indescribable. To be appreciated they must be witnessed. The horrors of the long drives, of the shipping-car, of the pens and slaughter-house, the cruelties of the dehumanized brutes in charge, the barbarity of the butcher, are beyond the reach of tongue or pen. If all would learn the truth of the horrible atrocities practiced upon the poor, patient brutes which form so large a portion of our daily food, many would turn from beefsteak, mutton chops, lamb stew and fowl with feelings of utter loathing. The shipment of stock, whether by rail or boat, is attended with the most horrible suffering. Many succumb to hunger, thirst, fatigue and ill-treatment before the voyage is half over. The report of the shipping companies reveal a shocking state of affairs. The poor animals, tortured beyond endurance, are driven to the slaughter-pens, and in this extremity of agony are converted into food for man. He takes into his system not only their disease produced by the brutal treatment they have undergone, but also the intense nervous strain from which they have suffered, which must have had a marked effect upon the physical being of the poor creature, and in eating the flesh man takes into his system the psychic elements of the agonized brute, and must of necessity suffer from the strain which so recently racked the flesh he is now making a part of his own body.

The Society for the Prevention of Cruelty to Animals

seems powerless to do anything to mitigate the horrors of this wholesale cruelty which is demanded by our refined civilization and permitted by our laws. This organization has, no doubt, done much toward ameliorating the conditions of animals which are used to draw burdens, domestic pets, etc. Their literature has had an elevating and refining effect upon those who have read it, and the general trend of the movement has been upward. They have as yet reached but a small majority of the people, and the good they might do is retarded by the stupidity of our law-makers and the blindness of those who are elected to execute the laws.

Habits of cruelty and daily sights of a disgusting nature so blunts the sensibilities of man, that he in time ceases to regard them with disgust. This gradual hardening of the whole nature is an interesting study, and proves to a certainty that the human family has been, and may be, brutalized by horrible scenes until all natural refinement is lost and he sinks lower than the brutes whose sufferings he despises. It proves, too, that man may elevate himself to a state so far removed from his present conditions that there will be no comparison between the two. It is to be sincerely hoped that before many years have passed the slaughter-houses may be abolished and the use of flesh as an article of food may be absolutely abandoned. Why should we allow the table to brutalize and degrade a certain class of men, whom, while society demands the product of their labor, hate their employment and shun their company as though they were beings of a different race?

Other evil effects of this wholesale consumption of meats are the various organic diseases which may be traced, either directly or indirectly, to the presence of parasitic diseases in the animals themselves, which are inducted into the system by eating their flesh, and often, too, disease is induced by decomposition of the flesh after death. It is a difficult matter to determine just at what point decomposition sets in, and few believe there is danger until the flesh is tainted, but certain it is that decomposition has begun some time before the odor was noticed. No process of salting, smoking or curing will render meat a safe article of diet. Only when every portion has become thoroughly heated will the parasites be destroyed. Hundreds of animals are thus diseased, and when we consider the immense quantities of half-cooked meats which are daily devoured, can we wonder that we are a nation of invalids?

It is not necessary that we enumerate the diseases which follow the use of meats. The tapeworm is familiar to many. The deadly effects of trichina has been often brought into notice of late years. The manifestations of this later disease are very similar to those of typhoid fever. No radical cure has yet been discovered, and, even if the patient escapes death, he rarely, if ever, recovers from the effects of the disease. It has been affirmed by some physicians that the consumption of diseased flesh is not attended with danger, but the proofs are so strong that it is useless to deny such a statement. It has been positively asserted by an eminent physician that the flesh and secretions, as well as the milk of animals affected with carbuncular disease, are so poisonous that those who handle them, as well as those who partake of them, are apt to contract the disease which are both painful and dangerous. The disease, usually takes the form either of an inflammation of the digestive canal or eruptions on different portions of the body. The vegetables upon which animals are pastured affects the flesh, and cases have been known where poisonous plants have caused the death of whole families who have feasted upon the meat of those animals poisoned by eating herbs, which contain sufficient poison to cause the death of those partaking of their flesh after their death.

It is thus seen that animal foods may engender many painful and dangerous diseases. The origin of a vast number of maladies which afflict many of the poorer people, who must use the very cheapest kind of foods, may also be traced to this habit of meat-eating. One of the most common diseases with which man is afflicted is scrofula. From it develops others more or less loathsome and fatal. It may be interesting to those to whom pork, lard and bacon seem a matter of necessity to know that the word itself is derived from *scrofa*—a sow.

Epilepsy is also augmented, if not wholly produced, by the use of animal foods, and those who have treated this disease most successfully are they who have advised the use of fruits, vegetables, cereals and plain, wholesome bread as a substitute for the raw flesh, rich soups and gravies, pastries and other abominations with which people are wont to gorge themselves. The fact that no lesions of other than an accidental nature have been found in the brains of those suffering from epilepsy seems ample proof that the disease is a functional disorder rather than an organic one.

It is quite unnecessary to enter into the task of discussing the many cases of cure, and the practically numberless cases of amelioration of disease and poverty which would undoubtedly result from the use of a fruit and vegetable diet. In this portion of my work I may

Henry S. Clubb

(See page 225.)

perhaps as well mention to my readers that I have practiced for years a strict regimen and am a striking example of the beneficial effects of a natural diet. Temperance in all things has been my motto, and I have found no cause to regret having discarded animal food. During the years that I have been engaged in the most exacting work, and have toiled almost day and night, I have found my physical health more nearly perfect and my mentality brighter, after adopting a natural diet, than ever before. To my simple habits of life as much as to anything else I attribute my perfect health and my capacity for accomplishing almost any amount of work I may desire.

With regard to epidemic infections, it has been demonstrated that those communities and those families which abstain from meats and from alcoholic stimulants are almost invariably enabled to withstand the diseases. During the frightful epidemics which sometimes sweep over a city or a whole country, the physicians and nurses who live simply and quietly, observing the laws of hygiene, and who do not become nervous and frightened, are enabled to resist the ravages of the plague, and to keep well and strong midst scenes which sicken men's souls.

Another interesting and certainly a most important point regarding this subject is that of economy, both individual and national.

The old saying "Where God sends mouths He sends food," is quite true—or, rather, would be true, did man but realize the fact that by obeying the dictates of loving Nature, and leading in all respects a natural life, he would find the earth yielding an abundance of all he might require for sustenance of himself and his family, for each child born into the world means a life to be supported and a pair of hands to till the soil and coax from the loam even more than is necessary for his own support.

At present much of our land is wasted. A given average of land devoted to the cultivation of wheat will support many times as many men as the same average would support if it were devoted to grazing or to raising grain with which to fatten stock for market. A great portion of the richest land in our country is devoted to pasturage, which, if devoted to a community living on vegetables, would sustain a large population; whereas at present it can sustain but a limited population. Thus it is obvious that our solution of the much-talked-of "Population Question" is found in the adoption of a vegetable diet. If this were done the inhabitants of the world might be increased almost indefinitely, for it is surprising how very small a portion of ground will fully satisfy the needs of any single individual.

Vegetables contain in themselves more nutrition than does the animal foods which they go to produce. This is a fact patent to all who will give the matter a moment's consideration. Why should vegetable products need to be converted to animal matter before we are able to obtain from them their real value? The theory that we cannot get from the vegetables themselves the same strength with which they endow the brute creation is a fallacy.

There is one branch of farming which in this country does not receive half the attention it should; that is the cultivation of fruits. If the farms in the United States were not so large, and were cultivated more thoroughly; if better conditions were induced, and a better stock of fruits and berries were planted, the results would be infinitely more satisfactory. The expense of so doing would be no greater than is the outlay for breeding fine animals, and the loss from climatic changes would be no more than is the loss of live stock from disease, etc. There can be no doubt that were public attention directed to this question, a wonderful impetus would be given to the cultivation of fruit, and especially to the finer varieties.

The gross ignorance of the poorer classes upon the value of foods can hardly be conceived. They blindly and stupidly follow the footsteps of their ancestors, and their obstancy in clinging to old ideas, their unwillingness to accept instructions upon the subject of health and morality is one of the most trying difficulties in the pathway of those who desire to better their conditions by decreasing their gross ignorance. They do not know that the way to gain health, which they covet, is simple; the way out of their superstition and ignorance less difficult than appears, even to those who are trying to raise them from a state of degradation to a condition of comparative comfort.

If asked, "What shall become of the animals? Will they not multiply so rapidly as to overrun the land? Will they not destroy natural products by which man strives to support himself?" Shall not our answer be that when man breeds animals for labor, and ceases to breed them for food, he can so regulate the numbers that there need be no surplus? It is true that artificial habits have for some time past disturbed the just balance of Nature, and many animals are now to be found in excess. We run no risk of being overrun by many animals for which we have no use, and which are allowed to propagate at will.

We must also remember that many of the animals are not natives of our soil; the pampered, unnatural lives they have so long led have made them susceptible to maladies almost as numerous as those which afflict mankind. The enfeebled brutes which to-day stand in their stalls or wander indolently through our pastures, are far removed from their progenitors. Relics all are they of a degenerate race, which would quickly disappear were they not cared for by man to pander to an acquired and vicious taste.

CHAPTER II.

HEALTH AND RIGHT LIVING.

DISEASE is not a foe to be subdued nor "cured" nor killed, but a kindly warning from Nature that her laws have been transgressed and that it is time you call a halt in your present methods of living and endeavor to conform to her teachings. Usually the first intimation of the fact that you have made a mistake is a slight indisposition which usually lasts but a few hours. These warnings often repeated, and as frequently disregarded, will result in organic trouble. It is but the voice of Nature, speaking in firm and gentle tones, telling you that you can not persist in your present course without serious consequences to yourself.

Disease is itself the defender and protector of the living organism, the life principle at war with an enemy, a penalty provided by Nature as a consequence of disobedience of organic law. It is a process of purification, an effort to remove foreign and offensive materials from the system and to repair the damages the vital machinery has sustained. Every attempt to cure or subdue disease with drug-poison is nothing more nor less than a war in the human constitution.

It is taught in all the books and schools of the drug-systems, that medicines have specific relations to the various parts, organs or structures of the living system; that they possess an inherent power to "elect" or "select" the part or organ on which to make an impression; and that, in virtue of this special "elective" or "selective" affinity, certain medicines act on the stomach, others on the bowels, others on the liver, others on the brain, others on the skin, others on the kidneys, etc. This absurd notion is the groundwork of the classification of the *materia medica* into emetics, cathartics, narcotics and nervines, diuretics, etc. Now the truth is exactly the contrary. So far from there being any such ability on the part of the dead, inert drug—any "special affinity" between a poison and living tissue—the relation between them is one of absolute and eternal antagonism. The drugs do not act at all. All the action is on the part of the living organism; and it ejects, rejects, casts out, expels as best it can by vomiting, purging, sweating, diuresis, etc., these drug-poisons; and the doctors have mistaken this warfare against their medicines for their actions on the living system.

The treatment of diseases with drugs ever was, now is, and always must be, uncertain and dangerous experimentation. It never was and never can be reduced to reliable, practical rules. An art is the application of the principles of a science to specific results; and a science is an arrangement of ascertained principles in their normal order and relations. These principles constitute the premises of the system which is made up of the science and the art. But in medicine, according to the philosophy of all the drug schools, every one of its fundamental premises is false; hence its science is false, and its practice must be false also.

On the contrary, the treatment of diseases with normal or hygienic agencies and materials is founded on the demonstrable laws of physiology, and reducible to fixed and unvariable rules of practice, and it affords the data for a true Medical Science and a successful Healing Art.

Wherever and by whomsoever this system is understood it is adopted. Just so fast as people become thoroughly acquainted with it, they abandon all the systems of drug-medication. Thenceforth they have very little need of the physician, and never patronize the quack. They will not be killed by regular, nor imposed upon by irregular physicians.

But an imperfect and superficial acquaintance with its fundamental principles causes many persons to err in the management of its agents and processes. The scarcity of properly educated Hygeio-Therapeutic physicians, and the incompetency and charlatanism of some who assume the title of Water-Cure doctors, have rendered it necessary, for the great majority who approve our system, to be their own physician. This every one can be if he will study the Science of Living as taught in our course of instructions and in our books upon this subject. Attention to the rules and principles herein stated and briefly illustrated will, I am confident, enable any person of ordinary tact and judgment to manage all ordinary maladies successfully, and to avoid doing any very serious injury in any case.

All that I have said, shall say, or can say against drug-medication, and in favor of the Hygienic system, is more than confirmed by the standard authors and teachers of the drug-system. I will give a few specimens of their testimonies on these points:

Said Professor Alex. H. Stephens, M. D., of the New York College of Physicians and Surgeons, in a lecture to the medical class: "The older physicians grow, the more skeptical they become of the virtues of medicine, and the more they are disposed to trust to the powers of Nature." Again: "Notwithstanding all of our boasted improvements, patients suffer as much as they did forty years ago." And again: "The reason medicine has advanced so slowly is because physicians have studied the writings of their predecessors instead of Nature."

The venerable Professor Jos. M. Smith, M. D., of the same school, testifies: "All medicines which enter the circulation poison the blood in the same manner as do the poisons that produce disease." Again: "Drugs do not cure disease; disease is always cured by the *vis medicatrix naturæ.*"

Said Professor E. S. Carr, M. D., of the New York University Medical School: "All drugs are more or less adulterated, and as not more than one physician in a hundred has sufficient knowledge in chemistry to detect impurities, the physician seldom knows just how much of a remedy he is prescribing," Again: "Mercury when administered in any form is taken into the circulation and carried to every tissue of the body. It often lodges in the bones, occasionally causing pain years after it is administered. I have often detected metallic mercury in the bones of patients who had been treated with this subtle, poisonous agent."

Professor S. St. John has said in unequivocal language, that "All medicines are poisonous;" while another physician of repute informs us that the more simple the treatment the better the result.

That some of the most noted among English physicians have but little faith in the efficacy of their remedies is proven by the following testimonials: "I have no faith in medicines," says a distinguished London physician; "the medical practice of our day is, at the best, a most uncertain and unsatisfactory system; it has neither philosophy nor common sense to commend it to confidence."

Another equally prominent practitioner says: "Gentlemen, ninety-nine out of a hundred medical facts are medical lies, and medical doctrines are, for the most part, stark, staring nonsense." And still another informs us that he is "incessantly led to make an apology for the instability of the theories and practice of physic. Those physicians generally become the most eminent who have most thoroughly emancipated themselves from the tyranny of the schools of medicine. Dissections daily convince us of our ignorance of disease, and cause us to blush at our prescriptions. What mischief have we not done under the belief of false facts and false theories? We have assisted in multiplying diseases; we have done more, we have increased their fatality."

It can not be denied that the present system of medicine is a burning shame to its professors, if indeed a series of vague and uncertain incongruities deserve to be called by that name. How rarely do our medicines do good! How often do they make our patients really worse! I fearlessly assert that in most cases the sufferer is better without a physician than with one. I have seen enough of the malpractice of my professional brethren to warrant the strong language I employ, and when we are told by those most prominent in the profession that "Assuredly the uncertain and most unsatisfactory art that we call medical science is no science at all, but a jumble of inconsistent opinions; of conclusions hastily and often incorrectly drawn; of facts misunderstood and perverted; of comparisons without analogy; of hypotheses without reason, and theories not only useless, but dangerous," and when another gravely assures us that "Some patients get well with the aid of medicine, more without it, and still more in spite of it," and another says with equal gravity and truthfulness that "Thousands are annually slaughtered in the quiet sick-room. Governments should at once either banish medical men, and proscribe their blundering art, or they should adopt some better means to protect the lives of the people than at present prevail, when they look far less after the practice of this dangerous profession, and the murders committed in it, than after the lowest trades," we can only say with one of their number: "Let us no longer wonder at the lamentable want of success which marks our practice, when there is scarcely a sound physiological principle among us. I hesitate not to declare, no matter how sorely I shall wound our vanity, that so gross is our ignorance of the real nature of the physiological disorder called disease, that it would, perhaps, be better to do nothing, and resign the complaint into the hands of nature, than to act, as we are frequently compelled to do, without knowing the why and the wherefore of our conduct at the obvious risk of hastening the end of our patient."

We can only wonder at the stupidity of the people who obediently and blindly swallow their vile stuffs, when assured by one of the most learned, "I may observe that of the whole number of fatal cases in infancy, a great proportion occur from the inappropriate or undue application of exhausting remedies," and by another that "Our actual information or knowledge of disease does not increase in proportion to our experimental practice. Every dose of medicine given is a blind experiment upon the vitality of the patient."

Despite their quackery they are forced to be honest as to results, and one perhaps of the most eminent has said: "I wish not to detract from the exalted profession to which I have the honor to belong, and which includes many of my warmest and most valued friends; yet it can not answer to my conscience to withhold the acknowledgment of my firm belief that the medical profession (with its prevailing mode of practice) is productive of vastly more evil than good, and were it absolutely abolished mankind would be infinitely the gainer." While another has thus unburdened his heart upon the subject: "The science of medicine is a barbarous jargon, and the effects of our medicine upon the human system is in the highest degree uncertain, except, indeed, that they have destroyed more lives than war, pestilence and famine combined." Still another has dared be honest by saying: "I declare as my conscientious conviction, founded on long experience and reflection, that if there was not a single physician, surgeon, man-midwife, chemist, apothecary, druggist nor drug on the face of the earth, there would be less sickness and less mortality than now prevail."

These extracts, which might very easily be extended so as to fill a large volume, shall conclude with the following confession and declaration deliberately adopted and recorded by the members of the National Medical Convention, representing the elite of the United States, held at St. Louis, Mo., a few years ago:

"It is wholly incontestable that there exists a widespreading dissatisfaction with what is called a regular or old allopathic system of medical practice. Multitudes of people in this country and in Europe express an utter want of confidence in their physicians and their physic. The cause is evident: erroneous theory, and springing from it, injurious, often—very often—Fatal Practice! Nothing will now subserve the absolute requisitions of an intelligent community but a medical doctrine grounded upon right reason, in harmony with, and avouched by the unerring laws of Nature and of the vital organism, and authenticated and confirmed by successful results."

It does not take a man with a divining rod to tell that the same general law which wards off disease is that by which it is cured, and that any knowledge which one has by means of which he can cure the sick is of no avail unless it includes a knowledge of means by which when a person is cured he may "stay cured," for it is palpably absurd to be shut up to the necessity of curing people constantly. Such a process is only a sham. In reality there is no cure. It is merely a labor quite unfruitful of benefits. Without health no man can be as great as Nature designs him to be. Philosophically speaking, as well as practically, Health is wealth. Without it the highest mental culture can never be attained, for in its absence the closest application of one's intellectual powers can not be exercised. Without health there can not be a thorough moral discipline or religious growth, for to the degree that one is sick does he lack control over his passions, as well as over the emotions of his higher nature. There is no curse on this earth to-day so heavy upon the people as the loss of health. It makes those who suffer it so dependent, so miserable, placing them on a charity list; where if they are not dependent for bread, they are for other little things, which, good in themselves, sicken and sadden when daily had under circumstances inevitably calculated to press home to those to whom they are tendered the conviction of their own nothingness.

Many of the ablest medical writers admit the impossibility of curing chronic diseases by medicine. Many more admit it in their daily practice, who have patients to whom they give no medicine, recommending instead means entirely hygienic. In acute diseases drug-doctors speculate and experiment more extensively, but in this department men occasionally arise who have the magnanimity to admit that they can calculate with no certainty on their medicines, these utterly failing under the most favorable combination of symptoms to exhibit those effects for the production of which it is supposed they are specifically adapted.

Thus calomel, opium, quinine, lobelia, belladonna, aconite, toxicodendron, arsenic, iodine, podophyllin, and the other poisons whose name is legion, and in whose tails there are a thousand stings, are daily given, and specific effects are looked for and calculated upon, but exactly opposite effects are produced.

Poison is poison everywhere, always is poison; its effects are modified by the vitality of the person taking it, not by any change undergone, and the worse possible use to which you can put a sick man is to give him medicine, that if you want to kill him you have only to drug him, and if you do not kill him, you will waste away all the greenness and freshness of his existence, so that life looks to him as desolate as a burned prairie.

Sick people who take these drug-poisons do appear to get well? With some there is sufficient life-force for Nature to draw upon, to so far expel the poison from the system that the sick, having a period of rest, recover. But no credit should be allowed the drugs taken, for they only hindered Nature in her efforts. All the credit should be ascribed to Nature's restorative powers.

Although the patient may recover, yet the powerful effort Nature was required to make to induce action to overcome the poison injured the constitution and shortened the life of the patient. There are many who do not die under the influence of drugs, but there are very many who are left useless wrecks, hopeless, gloomy, and miserable sufferers, a burden to themselves and to society.

If those who take these drugs were alone the sufferers, then the evil would not be as great. But parents not only sin against themselves in swallowing drug-poisons, they sin against their children. The vitiated state of their blood, the poison distributing throughout the system, the broken constitution, and various drug-diseases, as the result of drug-poisons, are transmitted to their offspring, and left them as a wretched inheritance, which is another great cause of the degeneracy of the race.

Physicians, by administering their drug-poisons, have done very much to increase the depreciation of the race physically, mentally and morally. Everywhere you may go you will see deformity, disease and imbecility, which in very many cases can be traced directly back to the drug-poisons administered by the hand of a doctor as a remedy for some of life's ills. The so-called remedy has fearfully proved itself to the patient, by stern suffering experience, to be far worse than the disease for which the drug was taken. All who possess common capabilities should understand the wants of their own system. The philosophy of health should compose one of the important studies for our children. It is all-important that the human organism be understood, and then intelligent men and women can be their own physicians. If the people would reason from cause to effect, and would follow the light which shines upon them, they would pursue a course which would insure health, and mortality would be far less. But the people are too willing to remain in inexcusable ignorance, and trust their bodies to the doctors, instead of having any special responsibility in the matter themselves.

Multitudes remain in inexcusable ignorance in regard to the laws of their being. They are wondering why our race is so feeble, and why so many die prematurely. Is there not a cause? Physicians who profess to understand the human organism prescribe for their patients, and even for their own dear children and their companions, slow poisons to break up disease or to cure slight indisposition. Surely they can not realize the evil of these things or they could not do thus. The effects of the poison may not be immediately perceived, but it is doing its work surely in the system, undermining the constitution and crippling Nature in her efforts. They are seeking to correct an evil, but produce a far greater one, which is often incurable. Those who are thus dealt with are constantly sick and constantly dosing. And yet, if you listen to their conversation, you will often hear them praising the drugs they have been using, and recommending their use to others because they have been benefited by their use. It would seem that to such as can reason from cause to effect, the sallow countenance, the continual complaints of ailments, and general prostration of those who claim to be benefited, would be sufficient proofs of the health-destroying influence of drugs. And many are so blinded they do not see that all the drugs they have taken have not cured them, but made them worse. The drug-invalid numbers one in the world, but is generally peevish, irritable, always sick, lingering out a miserable existence, and seems to live only to call into constant exercise the patience of others. Poisonous drugs have not killed them outright, for Nature is loth to give up her hold on life; she is unwilling to cease her struggles; yet these drug-takers are never well.

The brains of thousands of men, and tens of thousands of children, have been debilitated and their minds clouded with thick mist, and, in many cases, totally darkened, by those powerful, life-killing drugs employed as healing agents. How many mothers, in order to make their little ones sleep, have blunted their moral sensibilities and rendered their intellect obtuse by dosing them with laudanum, "cordial" and other medicines! If men would observe the laws of life and health they would never require medicine, and in most cases where they take it they would do better without it if they began in season to practice abstinence and not carelessly and ignorantly augment the disease. And if our physicians instead of confining themselves to the cure of diseases would lecture and inform the people how to preserve their health, though they might make less money they would save suffering humanity a vast amount of misery and premature death. "An ounce of prevention is worth a pound of cure." But owing to the bad organization of society men have no time to attend to their health, in consequence of which the violated laws of Nature compel them to find time to

Alice B. Stockham.

(See page 206.)

be sick, and to die sooner than they otherwise would.

When we have discarded medicines and have ceased to look to the physician for relief from illness, which is the result of our stupidity and wilful disobedience, and listen to Nature's voice, heed her loving commands and entertain a right mental attitude toward our bodies and toward the whole world we will grow well. The power of thought in itself is as potent a power in the cure and preservation of our physical beings as it is for creating desirable or undesirable worldly conditions.

The command "Be ye perfect," is based upon the understanding of human capabilities. He who within his soul entertains high ideals of human life, who recognizes only the divinity working in all, and through all; whose heart is pure, and whose being is flooded with knowledge of his own individual greatness, will express in his life the thoughts that dominate his mind.

Health is harmony. Disease is discord. Health is the outward expression of an inward content, a universal, all-prevailing peace. Disease is but the expression of inward inquietude, of mental disturbance, of inharmonious conditions between man and Nature. As the mental idea is always expressed in the physical life, health can be found and kept only by steadily adhering to the wise rule of contemplating only what we desire to see materialized. Health is the normal condition of every creature, and is obtainable by all who desire it. No one *need* be sick, but all can be if they will, and constantly hold in their minds thoughts of sickness, disease and pain.

There is no need for man to go to doctors for poisonous drugs to heal sickness, when by resorting to the simple, natural agencies so near at hand, he could perfect his health and, by observation of the laws, keep it in a state of perfection. Those who patronize doctors least are always in better health than those who are constantly under medical treatment and who take every new medicine placed upon the market.

Everyone knows that hot drinks, a vapor bath and a good sleep will cure a cold or a troublesome cough, where drugs would fail to produce the least result, unless possibly to make the sufferer worse by adding to his discomfort. Is it necessary then that one consult a physician and take his medicines if by the use of so simple a thing as a bath and a cup of pure water he can successfully deal with the case at home without the use of any harmful agency, and without the expense of advice?

Fevers and many other diseases for which a physician's services are usually considered necessary will respond as readily to simple, natural methods of treatment, and often organic troubles may be arrested by beginning in time. In fact we do not know of any disease which may not be successfully treated at home by common sense methods. There is a great deal more good in grandmother's remedies than in all the pills and powders of the doctor. There is vastly more virtue in a perfect diet, perfect cleanliness and an equable temper than in all the lotions and washes of the modern "beauty doctor."

Keep the body clean, the blood pure and in good circulation, thus giving vigor and tone to the whole system. Keep the feet warm and the head cool, wear proper clothing and protect the body well from all climatic changes. Do not be afraid to work, for it is the panacea for all kinds of miserable aches and pains. He who has nothing to do is unfortunate indeed, while he who has plenty of work has nothing of which to complain.

Advice as to natural methods of living and of attaining to the highest state of bodily health and mental perfection is given to our students in Personal Magnetism, and we have in this way been the means of benefiting many and bringing them to a full realization of the heights and depths of life in its truest sense. The true spiritual life wherein is all perfection, all contentment is indeed a life worth living. Like health it can not be bought and sold, but must be developed.

Disease may all be put in the same category and called abnormal conditions, created by wrong methods of living. Doctors are fond of classifying diseases and giving them high-sounding names which will appeal to the superstitious and make them think they have an interesting complication. It is by thus catering to the imagination and the emotional nature of man that they are enabled to keep up their practice and accumulate their immense fortunes.

The only way to keep in a healthy condition is to keep clean inwardly and outwardly. When the system is laden with waste matter, any slight irritation of an organ may cause a slight local trouble which, unless properly treated, develops into real organic disease. This is, of course, avoidable if one will begin in time to practice the rules of health and conform his life to Nature's highest laws. An English physician in a work on Health says: "Persons who inquire into the hygienic treatment of diseases are struck with the similarity of treatment which I advise, no matter what the complaint. All are advised not to smoke nor drink stimulants; to eat plain food in moderation; to have bread made of whole wheat flour instead of white; to exercise daily; to breathe fresh air always, and to keep their pores open. The four chief laws of life are: simple food, proper in quantity and quality and having a sufficiency of bulk to cause daily laxation; fresh air to purify the blood and keep up the heat of the body; exercise to keep the muscles in order and open the pores, and bathing to free the skin from dirt and to allow waste material to be the more easily thrown out. Of all the persons who visit me each week not one does exactly right, while many are wasting life and energy by using alcohol and tobacco. The majority are eating more than their bodies require; the excess food causes plethora, low spirits, bilious attacks, headaches, noises in the head, specks before the eyes, pain in the back, along the neck and behind the eyes; failing sight, loss of bodily and mental power, etc. Some are eating wrong foods, such as white bread, peeled potatoes, meat, or not taking enough fresh fruits and vegetables. These suffer from constipation, backache, piles, varicose veins, heavy, dull feeling and loss of energy. Some eat too fatty or greasy foods and subject themselves to heartburn, acidity of the stomach and skin eruptions. Foods made sweet by added sugar cause wind in the stomach, sleepy feelings, want of energy, skin eruptions, flushes of heat, etc. Some do not take enough exercise; these suffer from cold hands and feet, poor circulation, chilblains in winter, red noses and depressed feelings. Those who do not breathe pure air may suffer from inflamed eyes, deafness, a spitting up of phlegm in the morning, morning and evening cough, cold in the head or nasal catarrh, pleurisy, bronchitis, inflammation of the lungs, or any respiratory complaint, even consumption itself, may follow from a weakly person frequently breathing bad air. Those whose skins are dirty are subject to fever and internal congestion or inflammation. If we add to the complaints the cases of sore throat, deafness, heart disease, indigestion, cancer, short sight and loss of mental and bodily power produced by tobacco, and the inflamed stomachs, hardened livers and kidneys, diseased hearts and brains, gout and rheumatism, with nervous disorders produced by beer, wines and spirits, we get a list containing most of the ailments which afflict humanity. The infectious fevers chiefly attack those whose systems are out of order; if these do attack fairly healthy persons, they run a short, mild course. Now my readers will understand why I usually give to all much the same kind of advice. As long as right conditions are not observed complaints will be common. It matters not to me from what a person suffers, he must live rightly before he can get well. From my point of view, I know not many diseases but only a wrong state of the system and various local symptoms. All treatment must be much the same, with local variations and some dietetic differences, whether a man has indigestion, rheumatism, deafness, dimness of sight, consumption, nervous disease or apoplexy. The dyspeptic has overtaxed his stomach, the consumptive his lungs, the sufferer from eczema has disordered blood, and so on. Knowing these facts, I can not understand why a doctor wants to use drugs. If we put the patient under proper conditions, he will get well of himself. From my standpoint all drugs are poisons, and instead of curing disease they actually set up other diseases, thus making the wearied system carry an additional load, and so lessening the chance of recovery. I write so confidently because I have successfully treated thousands of patients suffering from different diseases, of mild and severe kinds, and have more cures and quicker recoveries than do the prescribers of medicines."

This is simply the experience of every practicing physician. If the people could be persuaded to give up their gross habits of life they would soon be able to give up their complaints and live a pure and healthful life. Those physicians who are teaching the people temperance, cleanliness and the value of natural remedies are the benefactors of the race. We can not too heartily applaud their efforts, nor too earnestly assist them in their work.

Of natural agencies in the cure of disease we place first fresh air. People can not be healthy physically nor morally if deprived of it. It is essential to the very soul of man. Next, sunshine. Plant life quickly droops and dies if this life-giving principle is taken from it, and no one can develop to the fullest if he does not absorb into his life the essence of existence. Next, water. Ah, how essential is this. What a hardship to be deprived of it, and how sinful it is that in a world three-fourths of which is water hundreds of civilized people should not have the privilege of bathing facilities.

As a therapeutic agent water is invaluable. Properly applied it will cure almost every ailment from which humanity suffers. The "water cure" was a fad for awhile, and has given place to newer fads, but as a medicine it is still as powerful as when it was advocated by fashionable physicians in order to hold a fashionable clientele as well as for its real value as a health-giving agent.

If the water cure was made a national remedy for many existing social diseases the state would be the better for it. People who are filthy can not be moral nor can they make good citizens. Our large cities are built upon such plans as would suggest that water, fresh air and sunshine were entirely unnecessary to the well being of the human race. It is no wonder that disease and crime are bred in the dark, dirty alleys and the low, filthy dives.

What is needed is broad streets, large squares and public parks, comfortable houses, well lighted and supplied with abundance of water. Those who have not studied the question of hygiene can not well appreciate the disadvantages under which the masses labor. So long as the lower classes are allowed to exist in such fearful conditions, so long will the body politic be affected by the ills produced by ignorance.

Health is the first condition of human happiness. Its importance to a nation or an individual can not be exaggerated. It is vigor, strength, development, beauty, serenity and fullness of life. It is the perfection of our earthly existence ; the fountain of all joys ; the spring of all blessings. It is the condition natural to man as to all organized beings, and just so far as he comes short of this condition he fails in the true end of existence.

Men become diseased by uncleanness, sloth, gluttony, drunkenness, debauchery; by breathing bad air; by unnatural methods of living. Physicians try to cure these diseases—not by cleanliness, pure air, pure diet, temperance and by a return to Nature generally, but by the most opposite methods. Have they by their horrible drugs, their hideous surgery and their bungling methods of administering even natural remedies done the human race good? Let our answer be in the attenuated forms and sallow faces; in the common lack of development and beauty; in the falling hair and rotting teeth; in failing sight and hearing; in the prevalent dyspepsia, hysteria and hypochondria; in racking rheumatisms; in torpid livers and diseased kidneys; in asthmas, consumptions and scrofulas; in the various fearful diseases which rack the human family, and make life, which should be grand and sweet, one long agony from birth to death.

Ask yourself if this is the natural condition of mankind, and your answer can not but be no, a thousand times no! If you reason from cause to effect, you can not fail to see how man has, by disobedience to the law, brought upon himself the ills from which he so grievously suffers. The law is good; it is disobedience to the law that brings the evils. If we again come into harmony with the law the evils disappear. It is simply cause and effect.

Miss Emma Todd Anderson,

Author "Health Foods and How to Prepare Them."

(See page 231.)

CHAPTER III.

DIET.

I AM often consulted as regards a diet which will afford perfect nourishment for the body and brain. The seeker for the higher life is confronted with grave difficulties the moment he starts upon his career. Accustomed as he is to a meat diet, and totally ignorant of the properties of the various foods, and careless as he has always been to the effect of diet upon his system, added to this the embarrassment of procuring the right kind of food as a substitute for meat, and the situation seems at first almost too difficult to be overcome. However, with grains and fruits we can build up and sustain our bodies to the very highest state of health. There is no dearth of food or of ways of serving it. In fact, the various ways in which it may be prepared and the natural delicacy of the grains and fruits are difficult to realize until one has endeavored faithfully to conform to a strictly natural diet.

The first essential is a good appetite. He who strives after the truer, holier life must not be governed by the fashionable dinner hour, must not listen for the voice of the breakfast bell to tell him that he is hungry, nor must he hearken to the call to tea unless he feels the need of refreshment. The question when to eat is quite as important as the one of what to eat. As a rule, I should say, eat only when you are hungry, when the system demands nourishment.

While I do not believe in the long-continued fasts as practiced by some of the Oriental teachers, I do believe that most people eat entirely too much. It is really surprising to know what a small quantity of food will fully support life and keep the mental powers in perfect working order. Some of the most brilliant men of the past have spent but a few pennies a day for foods, and their works come down to us glowing with beauty, living thoughts for all ages, pure inspired gems from the pens of men who lived clean and holy lives. To-day the men who stand at the head of affairs and shape the destinies of our Nations are not they who sit longest at the banquet and tarry long over the wine, but the men of simple, frugal habits, who care too much for their physical natures to abuse them, and value too highly their mental powers to becloud them with gross dishes and the fermented juices of grape and grain.

The terrible and degrading belief that the body is dependent upon animal food for nourishment is most difficult to eradicate from the minds of those who have been reared and educated in this way of thinking. The effects of a pure diet can not be overestimated, nor can they be conceived by any one until he has personally experienced them. The physical perfection which results and the mental growth is reward enough of itself; the higher, truer, sweeter conception of life, and the clearer spiritual visions which come to one who has given up carnal things, are indeed worth any sacrifice which may be necessary to attain them. But why talk of giving up flesh food as if it were a sacrifice? Surely one has but to go out into the pastures and watch the cattle quietly feeding upon the grass, to see the lambs at play beside the brook, to hear the call of the fowl and the song of the bird in the lane, to go from there to the fattening-pen, from thence to the stock-yards and shipping docks, to the slaughter-pens and the markets, to have all desire for meat taken from him. If this is not sufficient, let him study medical works, diseases and their causes, let him look upon humanity as it is to-day, and study the causes which have been at work to produce the terrible effects which everywhere confront us. On all hands we find sickness, drunkenness, disease, poverty, vice and suffering. Everywhere men and women are laboring earnestly to ameliorate the conditions which drag men down to the level of brutes, but they are working blindly, ignorantly and without effect. Nor can they hope to accomplish much until they have dug up the root of the tree of evil, from whose branches they gather sorrow and despair for the millions of to-day and the millions yet unborn.

When shall I eat? you ask. Eat when you are hungry, not before; never when you are greatly fatigued; and always eat but little when you expect to exercise violently or have some difficult mental problem to master. Do not eat a hearty meal just before retiring at night, but, on the other hand, do not go to bed hungry. A glass of hot water and one or two biscuits eaten very slowly will satisfy the cravings of the appetite and induce pleasant sleep. Never partake of food when laboring under great mental excitement, for the blood being drawn from the stomach will retard digestion. Never eat much when feeling ill, for when the system is disturbed it is a sign that there is something wrong, that Nature is endeavoring to correct some physical derangement, and she should be allowed an opportunity of doing so. When troubled with headache or feeling of biliousness, let your meals consist of fruits, bread and a glass of lemonade until the system has been thoroughly cleansed.

Regarding what to eat, no positive rules can be laid down for all, except this one, that you abstain from the use of flesh, for while one man may find cabbage a delicacy, it may be to another rank poison, and while one man may feast upon potatoes and parsnips with intense relish, the sight of them may be offensive to another. Every person should study his own individual case, and eat only such things and in such quantities as best agree with him. No one need be at a loss to find plenty with which to satisfy his appetite after he has given up the use of meat, for Nature has spread with lavish hand her table for her children. The golden grains may be converted into bread, her fruits, nuts and berries grow in every clime, and the ingenuity of man brings them all to our door and makes them palatable.

To those who desire to live more in accordance with the laws of Nature, we would suggest the following menu for a week. The bill of fare may be varied, and indeed should be. Nature did not intend that we should restrict ourselves in diet when she gave us so many kinds of fruits for food:

It is usually advisable to drink a cup of hot water half an hour before each meal. If there is constipation or biliousness, a little lemon juice should be added to the water. An excellent drink is made by pouring boiling water over oat-meal. The water will be rich, creamy and foamy. The same meal may be used several times. It should be drunk quite hot, and several minutes before eating. Rice-water is another delicious drink. The juice of an orange added to a glass of water is very refreshing and particularly delightful. The unfermented juice of the grape and the apple, if taken in small quantities, is sometimes beneficial. None but the purest water should be taken into the system. If this can not be obtained from natural sources, every particle of water used for drinking should be boiled or filtered. Water should never be left standing in a metal vessel; a vessel of earthenware is much more desirable. When water is boiled for drinking purposes, it is sufficiently purified upon coming to a boil.

No liquids should be used during meals, nor for an hour or more after eating. The hot water before eating will excite the glands of the stomach to action, and cause the accumulation to pass off into the bowels, thus leaving it clean and ready to begin work upon the food as soon as it reaches it. The secretory organs of the bowels are stimulated and the excretions passed off in a perfectly natural manner. It is estimated that it takes three minutes for water to pass through the circulation; it will thus be seen that hot water is a powerful therapeutic agent, as it opens the pores of the skin, quickens the circulation and throws off many of the impurities which would otherwise clog the system and perhaps cause serious illness.

The person who desires to perfect his physical powers must abjure coffee, tea, chocolate, milk and fermented liquors of every description, and must partake only of the drink which the gods provide—water—mixed sometimes with the innocent and harmless distillation of fruit, berries and grain.

The subject of bread, too, is an important one. The refined white flour with which the modern miller furnishes us is an abomination. No wonder the majority of those who make it the staff of life are thin, bloodless creatures, with just enough life in them to enable them to creep about looking like ghosts from another world. All the life-giving properties have been eliminated, and there is nothing left but starch, which contains but a small portion of the principles necessary for the sustenance of life. Wheaten bread made of whole wheat is a perfect bread, and has within it elements of nutrition for almost every part of the human organism. Few people realize what a perfect article of food corn bread is. It is considered as fit only for the very poor and the humble toiler, when it is indeed fit for the table of a king. When properly made it is as sweet and wholesome as one could wish. It has a peculiar, rich, nutty flavor that is delicious. Breads should never be eaten while hot nor while very fresh, but must be allowed to grow quite "stale" before it is digestible. One who has not learned how to eat and how to extract from an article of food the rich flavor which even the simplest dishes may contain is incompetent to judge upon the subject. A piece of cold, stale bread full of the natural oils of the original grain, a handful of nuts and dates, together with a bit of fresh, ripe fruit, may yield more enjoyment to the wise man than the most sumptuous banquet would afford the glutton.

However, to return to our previous question of a menu for a week.

Breakfast may consist of Shredded Wheat Biscuits, which, by the way, are most excellent, as all who have used them are quite ready to testify. They can be obtained from almost any grocer or direct from the wholesalers. Fresh, perfectly ripe fruits should always form a part of the morning meal—berries, in their season—oranges, bananas, peaches or any sort of fruit which happens to be in the market and which may tempt the appetite. For luncheon, wheaten bread, ripe apples, baked potatoes, and for dinner, corn bread, stewed squash, potatoes boiled with the skins on, baked apples and a slice of whole wheat bread, with figs and nuts for dessert. Breakfast on the second day may consist of toast made of rye bread slightly moistened with warm water, fresh fruit and fig sandwich. Luncheon, baked apples, baked potatoes, sliced tomatoes, celery, nuts and figs. Dinner, French fried potatoes, fried in vegetable oil or cocoanut butter, cabbage, sliced oranges, chipped apples, corn bread, nuts and fig butter. Breakfast, wheaten bread, fruit, sliced bananas. Luncheon, baked beans, corn bread, baked squash, peaches and a slice of wheaten bread, with lemon butter for dessert. Dinner, sweet potatoes, baked apples, chopped lettuce, onions, sliced beets, celery, toast of rye bread with dressing, washed figs, oranges and nuts. Breakfast, Shredded Wheat Biscuits with dressing, French fried potatoes, steamed asparagus, date butter and wheaten bread. Luncheon, potatoes boiled with the skins on, boiled cabbage, baked parsnips, peas, chipped apples with sauce, wheaten bread, baked bananas, dates and nuts. Dinner, wheaten bread, baked potatoes, baked apples, stewed squash, onions, lettuce, radishes, ripe apples, oranges, fig butter, nuts. Breakfast, toast wheaten bread with dressing, peaches, fig sandwich. Luncheon, sweet potatoes, hominy, stewed peas, steamed asparagus, corn bread, baked bananas, pears, nuts, date butter and bread. Dinner, boiled potatoes, corn, baked beans, boiled cabbage, rye bread, stewed prunes, rice, slice of wheaten bread, with figs, nuts and peaches for dessert. Breakfast, Shredded Wheat Biscuit, baked bananas, French fried potatoes, fruit. Luncheon, baked apples, baked potatoes, rye bread, steamed asparagus, boiled onions, beets; for dessert, wheaten bread, oranges, nuts, ripe apples. Dinner, boiled beans, boiled potatoes, cabbage, chipped apples with sauce, rye bread, peas, fig sandwich and sliced bananas. Breakfast, wheaten bread toast with dressing, wheaten bread with fig butter, baked apples, peaches or other ripe fruit. Luncheon, boiled potatoes, fried apples, corn, asparagus, peas, boiled onions, corn bread, fig sandwich, creamed bananas, wheaten bread and date butter. Dinner, French fried potatoes, baked beans, onions, lettuce, radishes, steamed asparagus, boiled potatoes, rye bread, baked bananas, with fig butter, dates, nuts, oranges and grapes for dessert.

Of course the reader will understand that the above are merely suggestions, to be changed as may be desired by the individual, and is given merely to show what a variety of vegetable products we have from which to choose, and that it is unnecessary to have a monotony in the bill of fare even for a week, and when we think of the rich products of the various seasons we are at no loss to plan dainty, tempting dishes for each succeeding meal.

There are, too, various cereals which can be introduced into the regular diet, and the different methods of serving fruits will give a still greater latitude for the housewife's skill. There are also many kinds of biscuits and whole-grain breads which are as delicious as they are wholesome.

Fresh raw fruits, perfectly ripe, and, if possible, just gathered, with bread and nuts, should form the chief diet. To this may be added figs, dates and other fruits, which, when dried, are preserved by their own sugars. Many people find this diet satisfactory, and seldom or never partake of cooked foods. There are others who add to it various grains, eaten raw or parched slightly. This is, of course, optional with the individual.

Our one dollar cook-book gives recipes for preparing many dishes which are both wholesome and palatable. It is pronounced by those who are capable of judging as the best, most practical and most artistic work of the sort ever put on the market. It has been prepared for those who wish to furnish their table with the choicest viands, and will be invaluable to the housewife, and especially to the mother who wishes to rear her children on plain and wholesome foods and in accordance with the highest hygienic and scientific laws.

The perfection of the art of cooking is nowhere more observable than among the Buddhists. They use nothing but vegetables, and yet with only these simple elements they have a great variety of dishes. No meat, butter, eggs, cheese, lard or milk enters into the preparation of any of their dishes. Nowhere will you find people of purer minds, loftier ideals and more steadfastness of purpose than they.

Food to be properly digested and assimilated should be carefully prepared, well cooked and daintily served. It then appeals to the eye and tempts the appetite. Properly prepared food is essential to the well-being of man.

Of all the farinaceous seeds the preference is justly given to wheat. When manufactured into bread, it has long received the appellation of the staff of life. So useful and necessary an article of diet is bread, that those nations which have no farinaceous seeds make something in imitation of or as a substitute for it. Notwithstanding, however, that wheat contains so large a proportion of nutritious matter, the ingenuity of man has succeeded in divesting so minute a seed of its best properties and grinding the remainder into a fine white flour.

Here is one great cause of so much sickness: White flour bread constipates and binds like plaster of Paris; it is the doctors' best friend, but the people's enemy.

White flour bread is robbed of the bran, that which lubricates the bowels. White flour breeds disease which ends in death. Whole wheat flour contains all the elements of the human body in the same equal proportions. A human being is an animated grain of wheat from heart to skin, from toe to brain.

Bread made of this flour is brown and sweet as a nut, beside which bakers' bread made of *superfine* flour tastes like trash. If you would be healthy and avoid doctors' bills use the former *only* with everything in it as Nature intended.

It can not be too strongly impressed upon the human race at large the utmost importance of this flour to the unborn babe. Mothers using it before and after childbirth will be kept in marvelous health, and many of the usual disorders will never appear. One fact alone is worth many times the cost of the flour, and it is this: thousands of men and women lose their teeth early in life from no other reason than because their mothers used the wrong diet before and after their birth. If this whole wheat flour is eaten liberally, a grand structure is built up in the babe's mouth and a foundation laid for teeth that will last a lifetime.

Of wheat flour there are three varieties: in the first

all the bran is separated; in the second, only the coarse, and in the third, none at all. The bread made of flour from which all the bran has been separated is that most commonly used, but bread made of flour from which none of the bran has been separated is the most wholesome. Bran operates as a stimulus to the intestinal canal by increasing its pèristaltic or worm-like motion, and for this reason always keeps the bowels open, thus obviating the tendency to costiveness produced by the use of bread made from superfine flour. The mucilage it contains also soothes the bowels, preventing any irritation that might result from the particles or scales of bran.

The ancients considered that bread most wholesome and nourishing which was made of flour retaining the whole of the bran that is contained in the wheat. Hence, the Greek wrestlers used no other bread than that made with coarse, unsifted flour.

Bread should be light, and none other should be eaten. There is no excuse admissible for heavy bread. Bread is better when at least twenty-four hours old.

Rice is very easily digested, but it is seldom made into bread; it is generally boiled or stewed. It probably nourishes a greater number of human beings than all other grains put together.

It is more than probable that the common maize or Indian corn will, in a great measure, supersede the use of wheat; the greatest evil, however, attending its use is that it is seldom eaten unless hot, and then with a large quantity of butter.

Oats and the various cereals upon the market, if cooked properly, are wholesome and very essential, but if improperly cooked give rise to indigestion.

Apples are perhaps the most wholesome and most valuable fruit we possess. When ripe they are easily digested and afford an agreeable addition to any meal. Pears, peaches, apricots, plums, cherries and all the different berries which grow so profusely in this climate are healthful.

Nuts as an article of diet are used much too sparingly. Instead of being used as desserts they should form one of the chief dishes. They are rich in nutritive oils and are very agreeable to the taste. They are easily digested if eaten properly. They can be prepared in soups and stews, where they take the place of meats. The oils obtained from them are made into various butters which can be used instead of animal fats for cooking purposes. In fact there is no more healthful article of diet than nuts however used.

A really perfect diet, one that will fully satisfy the cravings of nature and keep the system in perfect condition, may consist of unleavened bread, nuts, fruits and pure water. These are all man really requires, and those who have limited themselves to this diet are perfectly satisfied, saying they have no desire whatever to return to the cooked vegetables which they consider quite as gross as meats and animal products.

That a natural diet is more conducive to the symmetrical and harmonious development of each and every part of the human body than animal food is undisputable. It is also a fact that flesh-meat is also decidedly more heating and stimulating than vegetable foods as it quickens the pulse, increases the heat of the skin, accelerates all the vital functions, hastens all the processes of assimilation and organization and renders them less complete and perfect, and, consequently, develops the body more rapidly and less symmetrically—exhausts the vital properties of the organs considerably faster, and wears out life sooner. Flesh-meat also causes a much greater concentration of nervous energy in the various organs through which it passes in all the successive processes of assimilation than natural food, and consequently leaves those organs more exhausted from the performance of their functions, and causes a greater abatement of the sensorial powers, develops and strengthens the animal passions and propensities and modifies the moral sentiments.

Plato declares that the springs of human conduct and moral worth depend principally on diet. Whether this be true or not, no one can deny that the health and mental powers of man depend almost entirely upon the food he eats. This has been too clearly demonstrated to admit of a doubt. As the body is built of the foods taken into the system, it must necessarily follow that it will be like the material of which it is composed. The mind is affected by the body to such a degree that it is like it, gross or refined, coarse or beautiful, even as the physical organism within which it dwells. The actions, which are the expressions of the mind, will, of course, partake of the physical nature, and thus we see how necessary it is to refrain from those things which are not in accord with true holiness even in our diet.

Many people never attain to the full measure of success which might be theirs did they not keep their sensibilities constantly benumbed by foods and drinks. A hot, heavy breakfast will cloud the mind and leave the system exhausted. After a disagreeable morning, the cause of which is usually unsuspected, a lunch equally indigestible is partaken of, and then a hearty dinner, followed by alcoholic beverages and tobacco. Such methods of eating are enough to wear out the physical organism of even a strong man in a short time.

In order to be in perfect mental condition it is necessary that the food be light and partaken of sparingly, that no intoxicants be used, and that plenty of exercise be taken to keep the organs of the body in good condition.

I know that thousands of persons in civic life are in the habit of partaking of hearty suppers just before retiring to rest—and I know too that by virtue of powerful constitutions, and perhaps much active out-door exercise, there is occasionally an individual among such people who enjoys a tolerable share of health, and attains to seventy or eighty years of age—but I also know that ninety-nine in a hundred of those who indulge in such practice are broken down and afflicted with chronic disease before they reach fifty years, and a large majority of them are in their graves before they are forty years old.

Next in importance to selection of foods is the method of eating. Those who do not properly chew what they eat and get it thoroughly mixed with saliva, swallow unprepared food, and, consequently, they do not readily experience a feeling of satisfaction and so eat too much. The excess of food eaten lies like a weight in the stomach, causes a dull and heavy sensation, and the person feels hungry although he may have already eaten too much. Those who are troubled with these symptoms should practice eating very slowly, mix the food well with saliva, and thus give rise to systemic satisfaction before they overcrowd the stomach. Never allow an ounce of food to pass into the stomach before it has been properly prepared by the teeth and salivary glands for the action of the digestive juices.

Here I must protest against eating and drinking at the same time, as such a practice is injurious. The custom of filling the mouth with fluid whilst we have food in it leads to imperfect mastication and insalivation, and what is in the mouth is swallowed before it is properly prepared for the stomach, causing indigestion and stomach troubles. The rule of eating must be to eat at meal-times, and drink either before or after. By obeying this rule we get into a proper mode of eating; neither do we delay digestion in the stomach, as fluids must be absorbed before the proper breaking up of foods can go on.

If foods are introduced into the stomach without being properly masticated, washed down with liquids and in excessive quantities, it is no wonder the human organism soon breaks down under the immense amount of labor required of it daily. We treat our horses and dogs well, and are careful that they are not overtasked in any way. We see to it that their food is suitable in quality and quantity, and that they have the best of hygienic treatment, but we give our own bodies no such thought, seemingly considering them incapable of wearing out or growing weak from abuse.

It would be advisable for all who have not already done so to familiarize themselves with the elements of physiology, and when they understand the body within which dwells the human mind, and know the beautiful workings of the laws which govern every portion of it, they will be more careful of it and endeavor to keep it in more perfect condition.

No one would expect to do good work with imperfect

or worn-out tools, and yet people wonder why they can not succeed, why they fail to accomplish as much as others, when their bodies are worn out, every nerve racked with pain, and the system in a constant state of rebellion against the hateful and unnatural methods of life. If you expect to use a machine intelligently, you first study the workings of every part of it, even to the minutest detail. You can not expect to use it unless you do so. You never realize your body needs the same earnest study as any other delicate machine which is built to serve an end. Unless it receives it the results are failure of the entire organism.

If the dyspeptic is tired of life, of constant suffering with the dreadful feeling of depression and melancholy, he can change it in a very short time by observing a few simple rules of diet, exercise and cleanliness. Let him eat only such things as agree with him, and those in moderate quantities. Let every mouthful of food be thoroughly masticated and prepared for the stomach before it is swallowed, for, as some one so truly remarked, "We have no teeth in our stomachs."

The dyspeptic should cultivate a cheerful disposition and a contented mind. Unhappiness is the cause of much of the physical suffering of the human race. Nature's ways are wise, and if you will follow her guidance you can not err. Man was intended to be both healthy and happy. He must get his physical and mental nature into accord, as neither can be in good working order if the other is out of harmony.

"Eat, drink and be merry," and you need not be worried with dyspepsia, if you eat the right kind of food, at the right time and in the right mood. Know what foods agree with you and what are best adapted to your

DR. A. O'LEARY.

(See page 197.)

manner of life, etc. Study yourself and endeavor to take proper care of the delicate mechanism with which you have been endowed by Nature. Do not think that you can neglect your body and escape the just penalty of your carelessness, for you can not. So sure as effect follows a cause, so sure must you reap that which you have sown.

You would think a person very foolish who, having a beautiful machine, delicately perfect in every part, would pour over it some kind of corroding acid or allow it to become clogged with dirt so that it was ruined; and yet that is what most people are doing with their bodies, and no one thinks much about it. Dangerous foods and drinks are forced into the system without any regard to its need, simply because it is the fashion to eat and drink so much at certain hours.

Comparatively few realize the importance of foods or think of their effect upon the system, but the readers of this book, having been shown some of the dire evils following the use of certain articles, will, I hope, study the human body and its needs, and endeavor to live a truer, simpler life, knowing that such a life is full of health and purity, happiness and success. There are, I know, many who need only be taught the truth, and they at once fall into line with those who are seeking the higher, nobler things of life. I trust there may be many such among my readers.

One of the most important considerations is cleanliness in preparing and cooking food. The store-room, the pantry and the kitchen should be the neatest and daintiest rooms in the house. Everything should be in perfect order, and all the utensils for cooking should be perfectly clean. Great care should be taken that no foul odors contaminate the food, and that fruits and vegetables are not allowed to decay. The vessels in which bread is kept should be guarded with vigilance, and the cloths and vessels themselves need frequent scalding and sunning.

Tinned fruits and vegetables should never be used if it is possible to avoid doing so. The metal which is absorbed by the contents of the can is in many cases absolutely poisonous, and, besides, the material used for canning is not usually of the best quality nor is it some times as fresh as it should be; then, too, the surroundings of a canning establishment are not always the most desirable. Dried fruits are even less wholesome than canned fruits. There are, of course, some firms who send out first-class fruits in cans, but it is almost impossible to get really good fruit in this way. The best and safest way is to procure the fruit direct from the grower; you can then depend upon its being good.

As to the proper kind of food for each individual, this will depend upon the person himself. No two people will like the same diet, and no two will require the same amount of food. The student will require a very different diet from the man who toils in the field or the mine. The man who uses his brain will be able to do his work upon a handful of nuts and a few ripe dates or figs when engaged in severe mental labor, and should he eat too heartily his brain will be dull and refuse to perform its task. The man in the field will eat heartily of his simple fare in order to restore the waste tissue of the body. Few will err upon the side of too great abstinence in either food or drink. More people die yearly of dyspepsia than of starvation.

He who would be perfect in physical health and mental power must study carefully the effects of diet upon his system, must eat nothing and drink nothing that will in any way interfere with the harmonious working of the whole nature. The mind and body are so interdependent that whatever affects one will affect the other. If the body is built up of gross material, and if the system is overcharged with matter, the mind will partake of its nature and be coarse, vulgar and sluggish in its processes. While if the material used for building up the system be pure and such as Nature requires, the outer man will reflect the inward purity, the mind will be clear and its operations perfectly logical and harmonious.

The method of thought and life will have a great effect upon the personal appearance, and the whole life and thought will be greatly influenced by the quality and quantity of the food used. No one can afford to be ignorant of the important relations existing between the physical and mental nature and of the influence of one upon the other.

Man is a complex being, and the manifold workings of his nature are difficult to understand. Wise men have devoted their lives to the study of his past, present and future, have earnestly sought to understand his true destiny. Thousands have been disappointed, and still the problem confronts us unanswered if not unanswerable. The old masters seem to come very near the truth, but their grand teachings, their noble ideals and their pearls of wisdom have been lost amongst the mass of superstitious rubbish with which they have been surrounded by modern teachers, until it is difficult to discriminate between the false and the true.

Man, has by yielding to the voice of passion, corrupted his nature until he has become a creature depraved, ignorant of his own dignity and nobility. He has ceased to recognize the divinity within his soul. Shall he continue in this course age by age, growing more vicious, more degraded, more weak and low, or shall he cultivate the dormant forces lying in the tomb of sensuality, not dead but sleeping, and attain the heights to which he is justly entitled? Nature, the great healer, the great loving mother of all, is ever ready to help us to regain our lost kingdom, and by conforming to her teachings, following her laws, listening humbly to her commands, we can in time re-establish the harmony which originally existed between her and mankind.

CHAPTER IV.

HIGH THINKING RESULTS IN UNIVERSAL SUCCESS.

EVERY true student of Nature fully understands the occult meaning of the words so often repeated by teachers of truth, "As a man thinketh, so is he." For centuries priest and preacher have, parrot-like, assured us that as men think so are they, but neither priest nor preacher has attempted to make clear to us the meaning of the words. It remained for one, touched by the divine flame which inspired his utterance to comprehend the depths of meaning contained in that terse sentence: "As a man thinketh so is he." How necessary then that he think right thoughts, thoughts of love and peace, of harmony and contentment. How necessary that the fount from which springs the force that shapes and controls our destiny be pure. "Ye can not gather grapes from thorns nor figs from thistles." So if the fruit be sweet the tree itself must be good. "As a man thinketh," then, deep in his heart, "so is he." If into his life comes sickness, pain, disappointment, failure, poverty, death, it must of necessity be true that in his own mind exist the conditions which bring these things to pass, for as a man is inwardly so is he outwardly. No man ever failed in life who set his heart to accomplish some one darling purpose, to conquer every difficulty, to overcome every obstacle. There was ever present in his soul the one thought, the one desire, the one overwhelming idea which submerged all others and created conditions which insured the fulfillment of his purpose. No doubts, no anxieties, no troublesome thoughts of possible failure haunted him. Sure of the possibilities of success, confident of his own powers, thoroughly reliant upon the invisible force which "moves mountains," and fully determined that nothing should prevent him from accomplishing that he desires, he went straight on. Can you not imagine a Washington, a Grant, a Napoleon, an Alexander so firm, so self-reliant, so certain of ultimate victory? Did fear of defeat disturb the heart of the conqueror of the world, and does dread of failure fill the heart of the scientist as he searches for the deeply hidden truths of Nature? Does not the farmer sow his grain and plant his seeds in full assurance of the future crop? Does not Nature herself teach us that we should expect fruitage from our labors? When the tiny seed falls into the earth it is moistened by falling dews, warmed by the sun and the life which lay hidden within springs into active being. The natural sequence is growth of the plant, then the bud, later the blossom, and after that the fruit. The end, the accomplishment of the destiny of that little seed, which has given by due course of Nature a thousand seeds as the result of its having followed without question as to the final issue the laws of its being. There has been no doubt, no hesitation, no hurry, but a calm certainty, a serene reliance upon the invisible laws which work without ceasing.

True it is but a senseless plant, but do we not learn some of our most sublime and beautiful lessons from watching the workings of natural laws in the lower kingdoms, and shall man not follow the teachings of even the humblest flower or the meanest creature if thereby he is made nobler, and is able to bring into his own life the principles of love, faith and hope, which will enable him to attain the true end of human existence? The blossoming of the simple flower should teach him that his own attitude toward Nature should be one of perfect love, even as the seed teaches trust and hope, in the future. So he should be willing to sow. His work and study is the tilling of the soil, his thought the golden grain, his hope, his trust, his love, the blossom that gladdens his heart, and the accomplishment of his purpose the perfect harvest of his dreams. Shall he allow the cold, slimy worms of doubt to feed upon the roots, or the ugly caterpillar pillage the luxurious foliage? Shall a loathsome insect eat out the heart of the blossom and cause it to wither? Then will he but reap as he has sown. If he sows thoughts of failure, so will his harvest be, if he has sown thoughts of success, they will spring up and bear fruit, some twenty, some sixty and some an hundred-fold.

It is indeed true that "thoughts are things," and that the outward life is but an expression of the inmost thought. Every one can prove this for himself, and after a little careful thought upon the subject, will understand how necessary that the secret thoughts which leave their impress upon the character and shine forth from the eyes, shall be true, noble and kind.

He will understand, too, how impossible it is for man to be a hypocrite, for sooner or later the power of thought will stamp upon actions, voice and features the true character of the real man, and he will stand forth known of all men as a pretender, a cheat, a liar. No man dare call himself honest when the shifting glance of the eye, the nervous twitching of the hands, the subtle, indescribable air of trickery in every act betrays the fact that deep in his soul lurks the demon of dishonesty.

There are many things to be considered in this connection. First the question of heredity demands our attention. Back of the man you are studying, and living again in his life, are the souls of dead and gone ancestors who have all had a part in forming his character and controlling his destiny. His early training and environments have left their imprint upon his mind, every one with whom he comes in contact exerts upon him an influence, and in a measure detracts from the real force of the individual, but stronger than all these, more potent for good or ill is the power of man's mental attitude toward himself, toward the world and toward humanity. Though he were the son of a monarch, and though within his veins flowed the blood of conquerors, if he hold in his own mind a low, debased opinion of himself, he will not rise beyond that which he conceives himself to be. Water will not rise above its level, neither will man rise above his own ideas of himself, but as water always seeks its own level, so will man rise to the heights conceived in his own mind of his true worth. A little self-conceit is a valuable thing, and it is well for a man to have a very good opinion of himself and of his abilities. He will rarely fall below his ideal self, and will, by attainment of the ideal, be able to conceive of and reach out toward a higher and better. Fortunately for the human race the old theory that an humble man was a truly wise man has been exploded, and men have come to realize the fact that the surest way of gaining universal respect is to respect yourself, the surest way of commanding full recognition of your true worth is to first recognize it yourself, for if you are ignorant of it no one else is likely to discover it.

Pray do not misunderstand me, and think I advise a foolish impertinence, a disgusting self-complacency, nor a patronizing indifference to the opinions of others, nor pray with thyself as the Pharisee of old, "Lord, I thank thee that I am not as other men are." The old philosopher who said, "Man, know thyself," was wise. It is the man who truly knows himself, who has sounded the depths and learned the breadths of his own individual soul, who comprehends his own powers, who withal is modest and unassuming, who goes on his way and does his life-work to the best of his ability, not troubling himself as to final results, but firm in the knowledge that all will be well in the end, who is the wise man and the true philosopher. He does not seek to force upon others a knowledge of his powers, for while he is conscious of them, he has no feeling of pride that it is so. All who come within his radius instinctively feel that they stand in the presence of royalty, that he is a man who can master circumstances, because he is complete master of himself. To them he is a king among men, a being of a superior order, whom ignorantly they look up to and worship, failing to realize that within themselves is the same deathless principle, the same matchless power. These are the men who rule, by whose hand millions are swayed, whose words are watched with breathless eagerness, and whose smiles or frowns make and unmake kings and kingdoms.

Is it not wise, then, that all understand themselves, know their resources and how best to use the powers which nature certainly intended them to use for their own benefit? Would you say it were well that a

Bismarck, a Gladstone or a Lincoln should undertake to manage empires, if they were not first able to control themselves, did not know their ability to meet any exigency which might confront them, did not fully appreciate their own power and feel that in the end their efforts would be crowned with success? Would you think it were wise to entrust a general with command of an army if he felt that he was incapable of rightly directing their movements upon the battlefield, and if he was assured that the end of all his efforts would be disastrous? You would call him a coward and place a better man in command, one who was sure of himself and who felt confident of success. Is it any more reasonable to think that the man who goes into business life with a miserable sense of his utter incapacity to successfully carry out his plans, and with a predetermination to fail, will control the stock-market and become one of the prominent men of the nation? Determine to succeed and success is yours. Determine to fail and you are generally sure to fail, for your thoughts will be constantly directed toward ultimate failure, and so sure as effect follows cause will you get exactly what you expect.

This is not a theory but a principle, clearly demonstrable and proven in the lives of those you see in daily business life, proven by your own life if you will but take the trouble to follow the processes which bring you success or failure, joy or sorrow. The millionaire, the drivelling idiot, the whining beggar and the dissatisfied tiller of the soil, the striking miner, no less than the minister, the doctor, the gambler and the lout, create the conditions which surround them, and make or mar their lives. It is the power of thought, the force of the human will which makes the world in which we live, for, "As a man thinketh, so is he."

It is not well to be content nor to be willing to stop when one has achieved a measure of success. He who earnestly desires to live well, to help others and to make the race better, purer and nobler, should never give up to sloth, for it is only by constant activity, constant thought that he is able to accomplish good. It is only by thought and activity that he can keep himself in the stream of human life. The man who ceases to think becomes narrow and prejudiced. He grows cold and ceases to attract. It is only by being in harmony with the thought-world of the day that a man retains his magnetic influence over others. He who lives always in a dead past, who has no sympathy with the present, could in no wise exercise a living power over men. He must be in sympathy with the age in which those whom he desires to influence are living. It is the present realities which enlist men's sympathies, and the wise man always cultivates the present.

While it is true that high thinking results in success, it is also true that base and degrading thoughts result in moral degeneracy, and in individual and universal failure. That man and that people whose minds and actions are dominated by impure thoughts are impure, and until the whole tenor of thought has been changed, remain in that state. The old question as to which was first, the goose or the egg, as the goose came from the egg and the egg from the goose, might be asked here, but no one who will give the subject a few moments of earnest attention can fail to see that base and impure thoughts will generate base and impure desires, and no one, after carefully studying the question, will deny that the thoughts which dominate the mind can fail to leave their imprint upon the moral and physical being. What make up life? Is it not little things light as air, unimportant we would call them? If the constant dripping of one drop of water will eventually wear away the hardest rock, and if the microscopic deposits of tiny creatures in the depths of the ocean will rear an island, upon whose shores will grow and blossom the fruits with which to nourish a whole nation who shall elect to make their home upon its broad surface, how can we deny the power of such subtle, unseen forces as the thoughts which flash like lightning through the brain to build from within the soul, to make us lovely and our lives truly happy, or to give us characters which are detestable and make of life a dismal failure?

All have noticed the change which soon takes place in the expression of a little child who is taken from a loving home where it has heard none but kind words and experienced none but the gentlest treatment, and placed midst those who are cross and hateful. How quickly the look of trusting confidence is changed to the glance of fear, and the charming, dimpled smiles are superseded by frowns and angry looks. It is quite as interesting to note also the change produced upon a child who has been reared among unpleasant surroundings, when the conditions are changed and it finds itself in an atmosphere of kindness and love. The quick frown, the shrill screams and the passionate temper are soon replaced by smiles and loving gestures. The thoughts which are passing through those baby minds are the cause of the change, and which make their character either lovely or hateful, their life either happy or miserable.

If, then, the power of thought so dominates the life of

man, and if it be indeed true that every passing sentiment leaves its impress upon his life, if he can by the force of these creatures of his brain mold his life as he chooses, how important that man knows how to think correctly, how needful that he understands how to use the power with which he is thus divinely endowed, and how necessary that it be employed for high and noble purposes. How essential, too, that man's thoughts be concentrated upon one thing, for it is by concentration of energy that the greatest good is wrought, that the most mighty purposes are accomplished. If the thoughts are wandering and the forces scattered, no very great results will follow. Success in any undertaking depends as much upon this as upon any other factor. There may be a high purpose, a firm determination to win, faith, hope and courage, but if the forces are not concentrated, if the thoughts of the operator be allowed to compass too broad a scope, the end can but be failure. The great trouble with most people is that they desire to succeed in too many lines, and eventually make a failure of all, when if they would devote to some one object the forces they are wasting upon many, they would have little cause to complain of failure in life. An Edison devotes to electricity his every thought, working with an intensity which would be impossible to give to a half-dozen pursuits. Horace Greeley gives to newspaper work the best of which he is capable. He is master in his chosen line, and so it is with all those who have achieved brilliant success in any particular line of work. Whatever profession you feel to be your best field of labor you should adopt; persevere in your determination to succeed in it, and allow nothing else to come between you and your heart's desire, then you will be able to give to it your best thoughts and the materialization of those thoughts will bring you sure success.

"As within, so without." As within the soul dwell thoughts of success, so without will form conditions which will not admit failure. Ah, if we could but tell all the world of the truths which lead to the highest happiness, could but teach all men the true secrets of life, could but help the thousands of struggling ones from the depths of despair into which they have been plunged by the forces within themselves, and from which it is possible to escape only by changing the whole current of individual thought. Man has but himself to blame if he does not attain success.

The question of food in its relation to the mind of man, its effect upon human life and human destiny is an important one. Ancient philosophers fully appreciated the value of a perfectly natural and wholesome diet upon the physical, mental and moral conditions; the necessity of a pure, simple, loving life. They themselves partook of only the choicest articles of food, those which come directly from the lavish hand of Nature, full of her sunshine, rich with her nectar. He preferred the luscious melon, the blushing peach, the dainty berry glowing in beauty and the grape bursting with the purple wine, to beastly roasts and the steaming pottage, the gross pastries and the strong drinks with which others endeavored to satisfy the cravings of the inner man. A simple brown bread, rich in elements necessary for the perfect nourishment of the system, was to unperverted taste sweeter than the finest bread produced by skilled artists in the culinary line. The blushing grape was more tempting to his eye than the most costly concoction from the king's table. The fruits, nuts and berries from orchard, forest and garden were to him a more perfect diet than were the carcasses of slain beasts, the vile stews of vegetables and meats and the fermented liquors of those who made "gods of their bellies." In their simplicity they scorned the drunken riots of the debaucheé; in their wisdom they walked with Nature and listened gravely to her counsel, and steadily followed her precepts.

Men who have studied the relations between food and mind, who have carefully observed the influence of certain articles of diet upon the mental state, and the resulting effect upon the physical and moral life, have no hesitancy in declaring that the more closely we adhere to the laws of Nature, the more pure will be the whole being. The intellectual life will be ennobled, the moral force will be increased and the physical health will be perfected.

No man with an imperfect physical nature can be morally healthy, for a sound mind dwells only within a sound body, a well-balanced intellectual life is necessary to a perfect moral development. Physical health can be attained only by correct habits of life, proper food, proper clothing, attention to cleanliness, daily exercise and a decided mental attitude, which precludes the haunting thought of possible illness. When one knows that he has violated no law of Nature, and that he has taken into his system nothing that will in any way disturb the normal functions, he has no fear of being sick, but goes on his way in full assurance that as the conditions surrounding him and within him are conducive to health, he will not become a victim of disease.

The man in perfect health, full of vitality and power, confident of his own ability, full of magnetic force, generated by the cultivation of those subtle, invisible powers which are so little understood, but which are so potent for good or evil, full of loving tender sympathy for his fellow-men, and earnest in his desire to help them to a higher plane of life, with an instructive knowledge of human nature, and dominated by an idea which demands the highest and best work of which he is capable, is the man who succeeds. It is also true that any one who is earnestly desirous of cultivating his forces and perfecting his every power can do so. The faculties lying dormant in the human soul can be developed and trained to respond to the highest demands upon the perfect knowledge of self and the ability to control the forces which dominate human affairs. Some men may stumble blindly to success, most men achieve it, and it is those who work patiently, hopefully, without fear of failure or doubt of their own ability, who most truly succeed.

As with individuals, so with nations; as is the power of one human soul, so the combined force of many men. High thinking makes individual success, and universal nobility will achieve universal success.

It is not the number of your houses, the wealth of your mines, the amount of your money in bank which marks you as a truly successful man, but the real force within, the ego, the I, which stamps upon all you do and all you say the imprint of your true worth. If you stand in your appointed place, filled with a light which radiates from within and inspires others to higher deeds, if you are self-conscious and self-confident, full of faith and hope and courage, making the most of present opportunities, and creating within you conditions which will not admit of failure, you can say in your heart, "I have achieved success."

MISS GRACE B. MOORE.

Publisher of "Social Culture and Laws of Life" and "Moore's Marvelous Memory Method."

(See page 222.)

You feel this to be true and you know, too, that unless you are well and strong, in full command of every force, you will never be able to accomplish your desires. You feel within you the continual warring of the elements of your nature, the battle between the higher and lower raging in continual conflict. You know that unless you can conquer your appetites they will conquer you, and your whole being rebels at the thought of being ever under the sway of passion, and yet you do not know how to cast off the bonds that hold you fast.

It has been my aim in all my writings and in my course of Private Lessons in Personal Magnetism to teach the reader and the student to recognize the fact that his whole life was one tangled web; that thought, and food, and associates all had their influence upon him; that nothing was to be considered as trivial or of little worth. Everything that comes into your life makes it better or worse.

Perhaps no one thing affects you more, however, than the views you hold regarding animal life and animal foods. Such subtile things are thoughts, such silent forces, forever building, building, never at rest, using always the material furnished them by the inner consciousness, making of the outward physical being an exact reproduction of the inner being.

You have a desire to be well and yet you are not, you desire happiness and have it not, you desire peace and contentment and it does not come to you. You wonder greatly and are very much troubled that your life should be so unhappy, and you feel that fate is extremely unkind to you, forgetting that you are yourself to blame for every condition of life.

You would be willing to change your life if it were possible, and wonder how it can be done. You would make almost any sacrifice were it necessary to attain your ambition. You are ever dissatisfied with self, ever longing for a higher life, ever asking how it may be attained. You are able to find no solution to the problem which so sorely puzzles you.

Of course you know that within you dwells all happiness or misery. Many mysterious agencies have been at work to bring about the conditions that are yours. The sins of past generations have been visited upon you in the form of appetites and tendencies. You suffer for those things you can not help, but you suffer far more for the things you can help if you will.

How can this be? You suffer for the degrading idea you hold toward your body. You do not care for it properly, you do not feed it properly, and in consequence your system is in an uncomfortable state, and your mind suffers in sympathy with it.

Your highest duty should be to keep your physical condition perfect. "A sound mind in a sound body" is the greatest blessing a man can possess. Without a sound mind you can not have a sound body, and without a sound body you can not possibly have a sound mind, for the two are inter-related, inter-dependent. Whatever tends to the improvement and development of one will perfect the other in like degree. Whatever debases and degrades one will equally debase and degrade the other. This is one of those mysterious laws of Nature which are so hard to explain, and yet are so relentlessly sure in their workings.

Health means first a clean body, pure without and within. Bathing, exercise and diet will accomplish this.

It means also a pure mind which will keep all the functions of the body in normal condition. In regard to the question of diet you need to be as careful about how much you eat as about what you eat. The body needs a certain amount of material to keep it in repair and in good working order, it needs no more, however. If too much fuel is added to the fire the blaze is extinguished; so, if too much food is put into the system the functions are disordered. If the supply is not diminished the consequences can not fail to be disastrous.

As we have already shown, the mind and the body are so closely related, the one serving the other, each dependent upon the other, that it is impossible to regard intelligently the healthful condition of either without considering the connections and influences of the other. Physical weakness tends to establish corresponding mental frailty. On the other hand, proper intellectual activity and a healthy condition and exercise of the feelings affect bodily functions favorably, while dissipation of mental energy, excitement of passion, mania of every description, disappointment and despondency tend to destroy not only the inherent forces of the mind, but to engender corresponding physical weakness.

How necessary it is then that the physical and mental conditions be kept in harmonious relations, one with the other. Anything that tends to disturb the functions of the body will also disturb the mind, and the imprint of that disturbance is left forever upon the soul of the individual. You can realize the truth of this more readily by remembering how greatly even a slight illness will affect you for days, perhaps weeks after. There is a disagreeable mental state which can not but powerfully impress the whole being.

Even thus are you, unconsciously, perhaps, affected by the slightest thing that comes into your life. Your breakfast, your dinner, the greeting you gave a friend, the flowers you wore, the letter you received, the mental attitude you have held toward the world, have all left upon your heart, your character, your whole future life, an impress that can never be erased, and have imprinted themselves upon your face in unmistakable characters that all the world may read. There is no escaping this inexorable law of cause and effect, this silent, invisible but powerful working of a natural force, shaping your life in accordance with unchanging and unchangeable dictates of a Supreme Power.

Those solemn and impressive words are ever thundering in our ears, our own divine possibilities, our own dreadful doom, "As a man thinketh, so is he." We can never escape them, go where you will. Always they are reminding us of our inherent power, always they chant of our weakness, "As ye think, so are ye." Thoughts are things, and everything that comes into our lives comes as an expression of thought long held in the mind. Everything that comes into our lives gives us new mental impressions, changes and shapes the force of our thought to a certain extent. In turn these new thoughts crystallize. We find that life is made up of thought and actions, the thoughts of to-day are to-morrow expressed in word or deed, the thoughts which this year dominate our mind are the living reality of the future.

If this be true, and it is true, we can not over-estimate the influence of a pure and noble life which to-day genders noble, loving thoughts, high ideals and determination of purpose to accomplish the true end of man's existence. We can not guard too closely the portals of that realm of thought whence issue those mighty mysterious forces, so wondrously potent in the shaping of human affairs. We can not be too careful in excluding from our daily life influences which tend to weaken our powers and to destroy the foundations we have laid for nobility of thought.

I feel this to be a theme of all others the most grand, the most sublime, the most potent in the affairs of men, and I believe that all who have read our literature and studied our course of lessons in Personal Magnetism fully appreciate the force of the thoughts here presented.

We have already touched upon the subject of food in its relation to the physical and psychic nature, and have endeavored to show how the whole man is made physically, mentally and spiritually pure by adopting a pure and natural diet. No one who has studied the subject, even superficially, will attempt to deny the influence of diet upon man. The greatest thinkers of all ages have been unanimous in their opinions that the highest forces of his nature are best developed and sustained by the use of natural products of the soil. Nowhere among the great masters of science and the great students of human nature do we find advocates of a gross animal diet. We have found that they who have contributed most liberally to the rich stores of human thought have been men of simple, frugal habits, who, eschewing the gross luxuries of the table and the enervating habits of languorous idleness, have labored unceasingly, lived plainly, almost meanly, have thought deeply, and have, as a result of their "plain living and high thinking," not only exercised over their own race and generation a powerful influence for good, but have done much to shape individual lives and mold the destinies of nations throughout succeeding ages. These are the men who have been a blessing to mankind, teaching them to live above the gross and sensual things of life, to admire truth, virtue and honor, to aspire to nobler lives and to be strong in their efforts to attain to an ideal state of existence. These are the men whose memory we revere and honor. Their names are household words, their lives excite our sincerest admiration and their example we feel is worthy our emulation. The history of the human race would be quite different did we not have the nobility of life taught to us by these self-sacrificing men of power and genius.

No man was ever truly great who was not supreme master of his own passions and desires, who was not able to command his own forces, who did not thoroughly understand his own nature and know how to control self. Many men, through force of circumstances, have become notorious, many have had "greatness thrust upon them" through accident or because of the uncontrollable impulse of the masses to go in herds, to follow a leader, though he be even more weak and wicked than themselves, but none ever achieved greatness who did not realize in his heart the innate nobility of his own character, the forces that dominated his life, and how best to make them serve him.

It is not by giving way to every impulse that sways you, by yielding blind obedience to every passion that dominates you, by weakly submitting to circumstances which at present surround you, that you will gain self-mastery, but by wisely controlling impulse and passion, by rising superior to the present, by creating circumstances which will be more congenial to you, by living so truly, so purely, so grandly, so nobly to-day that to-morrow is rendered more sublime by thoughts of love; by words of comfort, by deeds of simple charity, making sunshine for the future; by being strong and brave, by giving to others of your own strength and courage, by determination and earnest perseverance accomplishing the duties of daily life, by self-control, by steadiness of purpose, by simple obedience to the divine laws of your being; this is the way to gain self-mastery; there is no other. All who would conquer must work patiently, earnestly, hopefully, never doubting of the success which will surely crown all earnest efforts.

Purity of soul is the result of purity of thought; purity of thought is the result of temperance, chastity, love. The man who is chaste in his life is necessarily temperate in his desires. He is not a glutton nor a drunkard, nor does he defile his person with the fumes of the noxious weed and his system with its deadly poisons. His hands are clean of the blood of slaughtered innocents, his soul of the blood of his fellow-men. He has no past errors to regret. Excesses have not quenched the fires of youth. No ghosts of a wasted past torture him with visions of what "might have been" nor mock him to scorn for his weakness. Erect in his superb manhood, strong in the consciousness of his own power, confident of his superiority over others less wise than himself, he stands a king among men, a creative force in the affairs of life, a dominating element in shaping important issues.

These are the men we admire and trust and revere. They may not be nobly born, but they are noble. Their origin may be humble, their name obscure, but they have that which is better than birth or lineage, grander than the empire of a king, more enduring than the annals of the mighty—innate nobility of soul, true grandeur of character.

Weakness is not wickedness, but weakness is a sin. The man who is weak, who is opposed to making an effort to overcome present environments, however hateful they may be, who is willing to be controlled by destiny, blown about by every wind of passion, knowing what is right but too idle to make an effort to bring his life up to the standard he deems desirable, will not fill any very important position in life. He is like a lazy drone among busy workers, a pretty painted butterfly, sipping the sweets of an hour of passion, careless of the future, disdainful of the past.

DUEL BETWEEN A BULL AND A TIGER.

See Page 97.

CHAPTER V.

CRUELTY RESULTING FROM DEPRAVED APPETITES.

WE have generally supposed that the horrible brutality of pagan days had passed away with the pomp and splendor of savagery. We look upon Mexico and her occasional bull fights with a feeling akin to pity. Spain we regard as quite barbarous, her pleasures a relic of another age which will pass away with the century. However, this belief seems hardly likely to be verified, as an article clipped from a late paper plainly indicates. We quote the article in full:

"Spanish cruelty is no more a fallacy now than it was a hundred years ago. If the rest of the world is advancing, Spain seems to be going backward. The sports which reddened the sand of the Roman arena are reviving in Spain, for bull fights have ceased to be bloody enough for the Spanish populace. A short time ago an Andalusian bull and a royal Bengal tiger fought a duel 'to a finish' in the presence of thousands of people in Madrid.

"The duel took place at what is known as the Plaza de Toros. The tiger, Cæsar, was a full grown animal belonging to an animal trainer named Spessardi. The brute was unusually ferocious—so vicious, indeed, that the trainer had found himself quite unable to master it. Under those circumstances the tiger's death would not be deplored, while the fact of his ferocity gave him additional value in the eyes of those who wished him to fight.

"The director of the Plaza had a cage seventeen yards square by four in height erected in the middle of the arena. This was to protect the spectators from any unexpected incursions by the tiger, and at the same time gave what was considered an ample field for the combat with the bull. Thousands of people filled the Plaza; in truth, it is said that no bull fight in years has attracted so great an audience as this unusual contest.

"Presently the bull and the tiger were brought into the arena. For a second the two animals faced one another; then the tiger, with a roar, bounded on the bull, avoiding the horns, and fixing both teeth and claws in the bull's flanks and belly. The bull remained still for a few seconds, and seemed to be sinking backward to the ground. The spectators thought this was the end of it, that the bull had at once succumbed to the tiger's fierce onslaught.

"Evidently the tiger thought the day was won, for it released its hold for a second. This was just what the bull had been waiting for, and a wild series of plunges followed, the result being that presently the tiger found itself on the ground.

"Before it had opportunity to regain its feet the bull was upon it and sank his horns into the tiger's side. The tiger screamed with pain and rage, as with a hoarse roar the bull tossed the great striped body into the air. Again and again were these tactics repeated, the bull occasionally throwing its adversary against the sides of the cage with terrible force before tossing it.

"After a little time the tiger ceased to resist or to make a sound, and finally fell to the ground and lay limp and seemingly lifeless. This was a new turn of affairs, and seemed to puzzle the bull. And Cæsar was really shamming, for when, a moment later, the bull thrust its black muzzle close to the tiger's head, it seemed hardly a second before that muzzle was compressed between the tiger's powerful jaws.

"A struggle beside which the others seemed mere play followed. Eventually the bull managed to release itself, after receiving shocking injuries, and stamped upon the tiger furiously. The bull followed the stamping by impaling the tiger on its horns.

"After a few moments of this the cage was opened, the bull rushed out and was goaded back to his stable. The tiger was found to have ribs broken beside having received a number of wounds from the horns of the bull. Nevertheless, it will survive, although its ferocity seems to have left it, and it is as mild mannered as a cub."

This shows that the old love for savage sport has not died out of the hearts of the Spanish people. The word "Spaniard" is synonymous with cruelty. The expression "he is as cruel as a Spaniard" has become current in every land where stories of the heartlessness and treachery of the Spanish nation have been told over and over again. The people of Spain long cultivated by every art the baser passions. They were a nation of meat-eaters and drunkards. All kinds of brutal sports were indulged in. The holidays were devoted to games of the most brutalizing character. The vast amphitheaters of the large cities attest to the magnificence of these spectacles. Ancient history gives most glowing descriptions of the royal banquets which followed. The civilized man of to-day reads with a feeling somewhat akin to horror accounts of the bull fights, but accepts without comment the spectacle of animal suffering in our own country. He glances over the columns of the shipping list without once thinking of the suffering consequent upon the long drives, the starving, the cruelties practiced by those in charge, the terrors of the voyage and the horrors of the slaughter pen. He learns how many thousands of pounds of beef and bacon and lard and sausage we have exported and of the immense amount consumed at home, without once remembering that all this quantity of animal food is obtained at the sacrifice of millions of innocent creatures, and without realizing that we must in turn suffer in many ways for this criminal murder of our fellow-creatures.

While we condemn heartily, the cruelty which permits the bull fight and the gladiatorial contest, we feel no thrill of horror when we read in the "sporting" columns of the newspapers how two brutal fellows are training for a contest in the "ring" which rivals anything ever produced by ancient Athens or Rome for brutality. We say "boxing" to-day is "scientific," that it has become an art and that any gentleman may witness such scenes and indeed indulge in them without in any way becoming demoralized. So were the sports of olden times conducted on scientific principles. The men who devoted their whole lives to the training of their physical powers were real artists in their line, and as we go back to those old heathens for the highest we have in poetry, painting and sculpture, so we must seek of them the secrets which made them so grandly strong and so wondrously skilled in their gladiatorial contests.

We do not shudder when we read of dog fights, cock fights and others of like nature, in our own country.

Neither do we consider it other than splendid sport when we learn of some sportsman who succeeded in bringing down numerous small game or in catching a good string of fish in one day, while all applaud our President if he happens to be a remarkably good shot and lucky hunter, and the keenest interest is felt in the number of birds and fishes any official may succeed in murdering during one day's outing.

The occasional bull fights of Spain and old Mexico can hardly be more shocking in detail than are the daily scenes enacted in the various slaughter-houses in our large cities. The suffering of one animal is not to be compared to the agonies to which thousands are subjected during shipping and while in the stock-yards. We make a great ado about the cruelties practiced upon one animal to satisfy a morbid sentiment in the thousands of spectators, and say nothing of the cruelties which are practiced upon the millions in order that man's abnormal appetite for flesh may be satisfied. The fact that thousands of the elite of the land are present at the gathering which gives them pleasure, while very few witness the scenes of butchery in the stock-yards, and know but little or nothing of them, makes no difference whatever, for the thousands gormandize upon the food thus prepared, and people are usually more brutalized by pandering to their physical appetites than by endeavoring to satisfy their mental demands for pleasure and excitement.

Were all people to adopt a natural diet, within two or three generations there would be no talk of war between nations. We would soon lose the feelings of hate, revenge and murder, which now dominate the human heart, and the affairs between countries would be conducted in a peaceful and loving manner instead of at the cannon's mouth. It is a sight to make the gods weep, to see nations calling themselves civilized, professing to believe in the "Universal Fatherhood of a Supreme Creator and in the brotherhood of man," building immense navies, fortifying their natural strongholds and gathering munitions of war. The trade in destructive apparatus is one of the most important, and the immense amount of money needed to keep up the various standing armies is no mean item of national expense.

The stronger countries of the world in their absurd endeavor to force their religion upon weak and defenseless peoples by the very forceful argument of gunpowder, is a spectacle at once amusing and sad, for it shows to what lengths the desire for power and the thirst for national fame will drive individuals. The missionary who goes into heathen lands to convert the inhabitants to his own way of thinking, with a Testament in one hand, a pistol in the other, and a knife in his belt, is not well prepared to civilize and refine the people.

Nor are national disagreements the worst feature. This cruelty to animals, and the use of flesh for food, exciting as it does the passions of the individual, make him quarrelsome and disagreeable in his own home, causing more or less unpleasantness, quarreling and fighting are the result. All the nobler instincts of both men and women are blunted, and their offspring, partaking of their nature, can but become more and more animalized with each succeeding generation. Farmers and stock-raisers take great pains with their fruits, vegetables and grains, their pigs, cows and horses; sowing only the finest of the wheat, carefully grafting and tending the choicest varieties of fruits and vegetables, breeding only the finest shaped, the gentlest tempered and most noble blooded among the animals, but we pay little or no attention to the breeding of a superior race of human beings. We seem to consider that this is in the hands of a power with which we have no right to interfere, and so leave everything to chance.

Truly we must suffer for our folly, for chance has seemed to take a special delight in propagating not the best nor even the mediocre, but the low, the vile and the diseased in most layish profusion. Upon all sides we see the workings of the great laws which govern the universe. How man, upright in himself, by the tempting voice of passion is brought down to a level with the brute creation, aye, and lower, for nowhere among the brutes are found such conditions as exist only too often among mankind. How swayed by appetite he commits all sorts of excesses and wastes his energies. How in ignorance of the causes which make life so unsatisfactory and fill the world with crime, misery and degradation, he stumbles blindly on attributing everything he can not understand to Providence, and wonders why life could not have been different, trusting in some dim future all things will be made plain to him, and that in the end eternal justice may be done. He does not realize that justice is being done now, to-day, nor that the divine law of compensation is being fulfilled every hour.

Men too often fail to relate the effect with the cause; in fact, they see only the effect, and suppose that because somewhere a cause exists, it must necessarily be as it is. They never look behind, never reason back to the forces which were set in motion and which combined to produce the events which have made themselves and their fellow creatures what they are. Neither do they realize that the forces which are to-day working in their lives will produce pleasure or pain, weal or woe.

It might seem a long way from the thought which prompts a man to strike the blow that brought a lamb to the ground weltering in its blood, to the cruel beating he gave his wife, and there might seem to be little or no relation between the piece of beef you ate for dinner and the coarse joke or the display of bad temper which destroyed the pleasure of the evening. You may fail to discern the effect of the generous bottle of wine upon your thoughts, or to note that when you have been most temperate in your habits you have been able to do your best work. Still the thought of depriving an innocent creature of its life in order that you might gratify your appetite, the eating of improper food and the excessive drinking of wine, have all had an effect upon the physical nature.

The effect upon national life is just the same as upon that of an individual, for the nation is but a collection of individuals, and it is a well known fact that people move in droves and blindly follow the dictates of fashion whatever they may be. So if meat-eating be fashionable, all men will be meat-eaters, and so far as they conform to hurtful lusts, so far will they become brutalized and degraded. A wave of thought sweeps over a nation, people grow to think alike. Most people accept without question whatever is the current opinion of the day. It is only now and then that a man is brave enough and original enough to think for himself. So if bull fighting, with all its attendant horrors, be the popular amusement of the day, all men will be found at the amphitheater and the thoughts which sway the heart of one, will sway the hearts of all.

Thus we see how a whole nation may become addicted to the use of meat, how a race may become a race of drunkards, and how popular ideas, which may have been advocated first by those who were actuated by purely selfish desires, may influence the destiny of nations. The life and thought of parents is, of course, reflected in the lives of their children. Left to chance, the race will soon degenerate, but it takes generations of culture to develop a noble race of people with pure thoughts and lofty ideals.

Culture improves the human race, neglect degenerates it. Good blood, right living, high thinking and a noble purpose in life are necessary conditions for those who hope to attain to perfect life and to really benefit the race. When we have grown wise enough to choose the father and mother of our children with the same degree of care that we select the dam and sire of the future colt or South-down sheep, and when we are careful to surround the mothers of men with the same kind and loving care which we bestow upon the mothers of the lower kingdom there will be a vast improvement in the race, and not until we do.

The people of primitive times seem to have had an idea of the importance of prenatal influence, and they also believed that the characteristics of the animal eaten were in some way partaken of by the individual. It is a well known fact that many of the savage races imagine that eating a portion of a foe, whom they had slain in battle, would give them the strength possessed by the man while living, and even among some of the Orientalists to-day the belief prevails that the man who slays many foes in battle is thereafter endowed with their physical prowess. This goes to prove that even in a savage state man had a conception of the great laws which control the destinies of man. They tried, in their simple faith, to obey them, and acting with the light they were able to command, sought to make themselves strong. Civilized man, though he has reduced the facts given to an exact science, is not yet so wise as were his savage progenitors, since he makes no effort to profit by his discoveries.

The effects of transmission of certain tendencies upon offspring can not be denied. We would say with Herbert Spencer, "If there has been no transmission of acquired character, there has been no evolution," and there has been no progress. No one will attempt to deny that there has been both progress and degeneration in the race. Families which were once honored members of society have, by the recklessness and wildness of some one of the family, been afflicted with a tendency for strong drink, and the proud old race has gone down; while another family, low in social scale, has by culture and development risen to prominence in the world and become a controlling force in national affairs. If such things are possible, and we see them exemplified in daily life, why are we not willing to make an attempt to conform to the laws which we can not fail to see must be for the benefit of the race? We talk much of reform, but we begin our work at the wrong end and after the mischief is all done, when if we began right we would save much of suffering and would accomplish more real good. If we would endeavor by following a natural diet to make our lives pure and clean, by plain living and high thinking to cultivate the nobler tendencies of the soul, to control the appetite and passions which if indulged degrade the life and place man

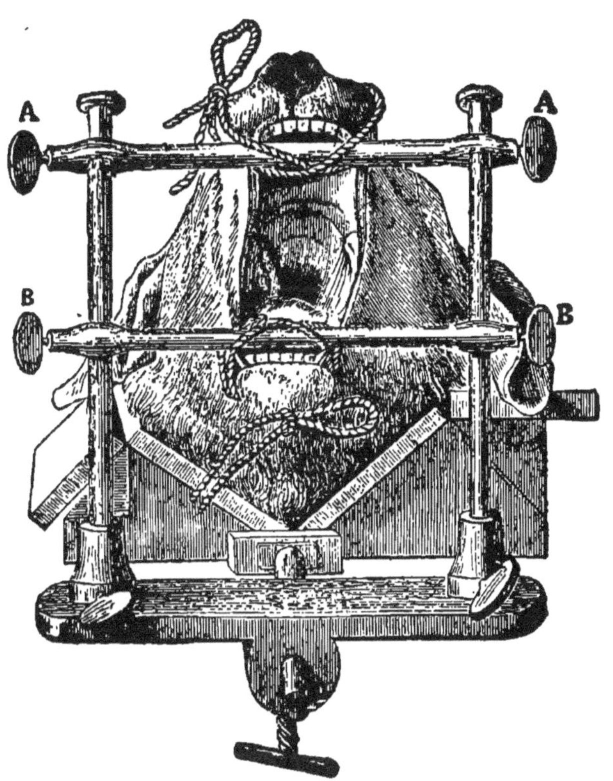

The mouth held open for hours, or at the experimenter's convenience. Imagine the misery of this condition.—From Bernard's *Physiologie Operatorie*, p. 137.

The illustrations used in this book picturing the horrors of vivisection are kindly furnished us by Mrs. Fairchild-Allen, Secretary of The Illinois Anti-Vivisection Society, and editor of *Our Fellow Creatures*, a monthly journal opposing the inhuman, unscientific and wholly unnecessary torture of animals.

on a level with the brutes, to allow nothing to come into the life which will in any way demoralize the forces which, if rightly directed, will make him but "little lower than the angels," if those foods which create disease and which inflame the desire for liquor and for other harmful indulgences were left out of our diet, man would in a short time become less cruel, scenes of domestic unhappiness would no longer disturb the serenity of the home, crimes of the most revolting nature would not stain the record of our country's pages, and war, with its attendant horrors, would be abolished. Men would cease to find pleasure in carnage. The cruelties of the slaughter-pen and the disgusting scenes of the packing-house would no longer harden the hearts and brutalize the natures of thousands of men. Children born into the world would no longer receive an heritage of disease, drunkenness and crime.

Surely it was never intended by Nature that man should fall so low, that he should become steeped in sensuality, that his whole nature should be base and mean, that he should revel in the slaughter of his fellow-creatures and indulge in all sorts of brutal practices. In no other department of organic life do we find that degeneration is the law of existence, but rather that progress from a lower to a higher state governs all matters. We find, too, that man is the only creature that is able to control his destiny, for, while he is controlled by the same great laws which govern the universe, he is able to so shape his thoughts and desires that his life is made in accordance with those desires, either high or low, either noble or base. The power of thought is the force which shapes our lives; the food we eat, the pleasures we enjoy, the attitude we hold toward every living creature, exercises an influence, makes us mild or vicious, loving or hateful. Every thought that flashes through the mind leaves an imprint upon the life and character of the individual. Is it not, then, important that the thoughts which so powerfully impress for all time shall be pure and beautiful and good, that the surroundings which call them forth shall be such as are conducive to high and noble life?

It is not in the shambles where bleating creatures stand trembling and submissive, not in the slaughter-pens where blood flows like water and the cries of murdered victims rend the air, nor in the market where dead bodies are exposed for sale, nor yet at the feast where smoking viands, rich in their own fats, heavy and gross and sensual in their very essence, are set out for the delectation of those whose natures have already been made coarse by indulgence; neither in the wine-room, where men go to quench the excessive thirst brought on by the stimulating effects of the strong meats of which they have partaken in abundance, that the highest and noblest virtues are cultivated. It is not in the chase nor in the field of battle that the truest depths of man's nature is stirred and that the best thoughts are generated, but in the peaceful pursuits of mankind among the grand scenes of Nature, in the cultivation of the soil, the development of natural resources, in contemplation of the beauties that everywhere surround us; it is in living in accordance with the laws which govern every department of Nature, in violating none. When man realizes the relation between cause and effect, and applies it to his own life; when he finds that in violating the law he but sins against himself, and that for every act of cruelty committed upon a creature more helpless than himself he must bear in his own soul the stain of a sin which can be wiped out only with long years of patient self-control and self-development, let us hope that he will cease to needlessly torture and slay, that he will satisfy his appetite with the products which Nature so bounteously offers, and that he will so govern his thoughts and desires that they shall no longer be his master, but he shall in his uprightness and strength rule his own kingdom.

A man, whether he be at peace with his kind or not, revels in a carnage of blood, the extent of which is endless, for it reaches round the world. Millions upon millions of people lust for blood, feast on flesh and sate their appetites by slaughter, and yet by taking the name and professing to emulate the example of the meek and lowly Christ, call themselves Christians. With uplifted hands, on bended knee, they pray to God for mercy, while near them can be heard the dying groans of helpless beasts, whose mangled flesh will deck the tables of their homes. At the bloody feast they will eat the dead remains of beings whose untutored lips could speak no pleading words, who could but yield reluctantly their life victims of savage inhumanity.

The eating of dead flesh has endowed the race of man with every brutish passion, filled the pages of history with war and painted the earth with blood. By it the purity of individuals and nations is sullied; by it the youths of ages have been debased in their instincts and made gross by their desires. A vast portion of the race has departed from natural simplicity and by perverting the uses of the blessings of earth has sated itself by the gratification of sensual desires, and prevented itself from the attainment of its true destiny.

There is implanted in the breast a natural repulsion for the carcasses of dead creatures, yet so perverted have the natural instincts of man become by custom, that when the carcasses or corpses of animals are cooked and set before them, they eat them with great gusto and apparent relish. The flesh of animals is but food which has been obtained from corpses containing the poisonous germs of corruption and death, while, on the contrary, fresh vegetation is live food, containing the seeds and germs of vitality. It is an undoubted fact, well sustained by scientific observation, that a diet of grain, fruits, nuts, etc., on account of its non-stimulating effect upon the animal passions, is favorable to purity of thought, and a harmonious and peaceful disposition. If the non-flesh eaters were to gain no other reward, this alone would be worth the effort of many generations to attain. While we, as a race, fondly arrogate to ourselves the perfection of refinement and ethical culture, it is a curious anomaly that we are willing to submit to and even to approve the conditions of cruelty involved in the slaughter and bloodshed of animals at which the highest sentiments of humanity should revolt with horror. The sickening scenes which inevitably surround the transit and slaughter of suffering, sentient beings is a travesty upon our arrogated civilization.

Mr. H. McChesney of Geneva, Ill., former Chairman of the Illinois State Board of Live Stock Commissions, is authority for the following plainly stated facts:

"During the season of 1888, persons supposed to be in the employ of the Butchers' Association purchased cattle diseased with Actinomycosis (Lumpy Jaw) at Chicago and other western cattle markets, and shipped them to eastern cities and paraded them through the

Animal suffering from Actinomycosis (Lumpy Jaw).

streets to exhibit to the citizens the character of the animals slaughtered at Chicago and shipped as dressed beef for their consumption. The Chicago market was destroyed, and business in the dressed beef trade so crippled the slaughterers were unable to sell their products because of the belief of consumers that the animals slaughtered were diseased. The slaughterers, commission men and officials of the Stock-yard Company urged the Board of Live Stock Commissioners to take action to prevent the slaughter and sale for food of all cattle diseased with Actinomycosis (Lumpy Jaw) that the market might be reclaimed and the slaughterers be enabled to sell their products in this country and throughout Europe.

"The Board was urged to employ John McDonald as its agent to seize all such diseased cattle, the slaughterers agreeing to pay an amount as compensation above that paid by the State sufficient to secure his services. On October 15, by direction of the Board, McDonald commenced the discharge of his duties by seizing and yarding, in pens set apart by the Yard Company, all the cattle supposed to be diseased with Actinomycosis (Lumpy Jaw), and such of them as the State or Assistant State Veterinarian found to be so diseased were held in quarantine till disposed of by the consignees for purposes other than for food. Later, at the request of the Committee of the Live Stock Exchange having charge of the disposition of cattle so diseased, a permit was granted them to make a contract for the slaughter and rendering of the carcasses into product not used for food.

"When, after great expense to the State and the untiring efforts of the Live Stock Commissioners, the Chicago market was reclaimed and all objection to the use of dressed beef products dispelled, one dealer became blind to the interest of the trade by the reflection of the almighty dollar, and shipped cattle diseased with Lumpy Jaw for slaughter and sale to the public, and because McDonald seized them he refused further payment for his services, saying he would not pay a man to seize his good cattle, thereby showing his anxiety to supply the public with beef from his Lumpy Jaw cattle.

"On the 22d of June, 1889, the State Veterinarian and McDonald seized a steer diseased with Lumpy Jaw purchased and shipped to Chicago for another dealer, who desired to slaughter the animal, but the Board would not permit him, and held the steer in quarantine seventy-two days, when the owners turned the animal over to the rendering company. This firm also refused further pay to McDonald for his services as per their agreement, and after thirteen months others interested liquidated their claim. These acts tend to show that the charge that cattle diseased with Lumpy Jaw were slaughtered in Chicago and shipped East for consumption was founded on fact, and would have justified the commissioners in withdrawing McDonald from the Stock

Yard and in leaving the Chicago cattle market in the fostering care of these corrupt dealers.

"The mental ability of these fellows will never enable them to comprehend the full extent of their obligation to the Agricultural Department of the Government for its assistance in furnishing the public with beef from cattle diseased with Lumpy Jaw by placing the Government meat inspection tag on the carcasses, thereby certifying to their healthfulness; nor are the beef-eating people of the country yet prepared to express their full measure of contempt for men who prostitute their official positions to a nefarious practice that subverts an Act of Congress and warps the integrity of the Government.

"That the preceding characterization may be clearly understood, I submit that on September 10, 1891, the city inspectors condemned four carcasses of beef in one slaughter-house because the animals were diseased with Lumpy Jaw. September 14 the city inspectors found several beef livers diseased with Actinomycosis on the killing floor of one slaughter-house. The Government Veterinary Inspector said 'such diseased livers were common in still-fed cattle, and that two or three could be found at each run of the killing.' He also said the carcasses of beef were fit for human food, and that it was only prejudice in the minds of the people who condemn them. R. W. Hickman, the chief meat inspector for the Government and veterinarian, said livers diseased with Actinomycosis are frequently found in still-fed cattle, but if he did not find the disease in other parts of the animals he would not condemn the carcasses. These carcasses were tagged with the Government meat inspection tag, and sold for human food.

Microscopal investigation showed the disease to be Actinomycosis (Lumpy Jaw)."

The following is testimony of some of the most eminent veterinary surgeons and physicians in this country regarding the danger of eating beef from cattle diseased with Actinomycosis:

"The flesh of an animal affected with Actinomycosis is not fit for human consumption, unless it has been first heated for a length of time to above 212 degrees Farenheit so as to destroy all organic life. The disease, once started in any part of the body, being liable to be extended to other organs through the cells; being carried by the blood, as well as by swallowing or inhalation, it follows that no part of the body can be considered quite safe when any part has become the seat of Actinomycosis. It must be understood that when one or more cells are starting to grow in a new tissue, there is at first no reason to suspect their presence."—*Dr. Law.*

"The flesh of animals affected with Actinomycosis is unfit for human food, because, first, of the presence of parasites, that may be lodged where the naked eye can not detect them. Second, because slightly diseased organs may often be sold to consumers through ignorance or greed. Third, because it takes more heat to kill the parasites than is often produced by the ordinary cooking of many people. Fourth, because it is possible that this parasite produces a poison of the character of Ptomaines, or some such principle."—*Dr. Paquin.*

"I believe the meat of Actinomycosis animals is entirely unfit for general use, as the introduction of the parasite and its possible growth is all that is desired to give rise to the disease. I think the fact of possible difficulty of discovering the same in all the tissues, and yet found in some parts of the body, ought to be sufficient reason to consider the whole as unfit for eating." — *Dr. Liautard.*

"The flesh of an animal with this disease is not fit for human consumption because of liability of transmitting the disease, and on the general principle that all diseased meat is unfit for human food." — *Dr. Knowles.*

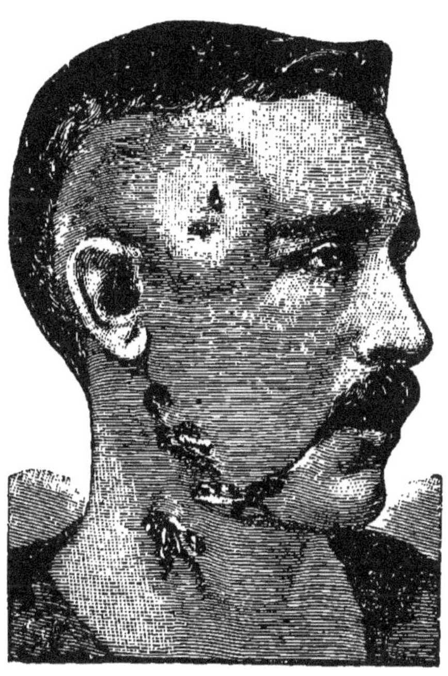

Case of Actinomycosis (Lumpy Jaw) Diagnosed by Prof. George A. Bodamer, M. D. B. S., Physician-in-Chief to the German Hospital, Philadelphia, Pa.

The opinions of these men and the accompanying cut should be enough to satisfy any reasoning person that the flesh of animals is dangerous even though it may have the appearance of being perfectly healthy, and may have been passed upon by the meat inspectors.

Prof. J. M. Byron, M. D., Director of the Bacteriological Laboratory of the University of the City of New York, has treated human beings diseased with Actinomycosis, and says it is fatal when the internal organs are invaded; that th " e is not so rare as was once supposed; and that to prevent its spread to human beings from the lower animals, it is absolutely necessary that an examination by intelligent men should be made of all animals at the slaughter-houses, and all found to be so diseased should be at once condemned.

Scientists who have made the disease Actinomycosis an especial study, testify that it is identical in animals and in man, and can be communicated to human beings by their eating the flesh from Lumpy Jaw cattle. "The fungus may attack any abraded surface, as the mouth, finger, stomach, intestines, lungs, etc., or it may attack the ribs, vertebræ or brain without any apparent connection with an external lesion." Inasmuch as the true character of this disease is of recent discovery, we can not know how many thousands of human beings have found untimely graves by their eating meat from Lumpy Jaw cattle.

When I consider that twelve cases of this disease in human beings have recently been treated in Chicago, I deem it just that the public should know the danger from eating beef from Lumpy Jaw cattle, and that it is sold for human food.

MRS. FAIRCHILD-ALLEN.

Founder of the Illinois Anti-Vivisection Society.

CHAPTER VI.

VIVISECTION.

ONE more terrible chapter on cruelty to animals, for, having said so much, we could hardly in justice to ourselves and to our readers, close this recital of horrors without giving a short treatise on that most terrible of all "arts" the art of Vivisection. "Man's inhumanity to man" has been a theme for writers and speakers for ages, but man's inhumanity to poor dumb brutes, upon whom he has exercised every species of cruelty that savage ingenuity has so far been able to devise, is a topic which has been passed over in silence by the great majority of those who are interested in the living questions of the day. Here and there one has arisen and has given to the world truths regarding the subject, but it is a brave man who will defy the lion in his den, the vivisector in his laboratory, who will stand boldly against the things in which public indifference and stupidity quietly acquiesce.

Science is the ruling passion of the age. No stick or stone, no phenomena in Nature, no fact in life escapes the keen eye, the strict analysis, the ready knife of the man whose soul is devoted to the study of natural law as is the soul of the lover to the mistress of his affections. The true scientist is calm, deliberate, heartless; he recks not of human or animal suffering. To him the world is one vast hall of learning, and spread about him are the inmost secrets of Nature awaiting but the touch of his hand to show him depths of knowledge as yet undreamed of. He cuts into the quivering flesh, sunders bones, arteries and ligaments, and notes the exquisite torture of the sufferer with all the delight that he would experience in dissecting a plant or analyzing a mineral or the formation of a bit of clay.

The question has been asked, "What is cruelty?" and like another propounded by the Roman Governor of old bids fair to remain unanswered for centuries to come. One who has perused the records of the various schools of physiology would perhaps say that the height of exquisite cruelty had been reached by these devotees of vivisection, but they are in turn confronted by the question, shall man, standing at the head of sentient animals, feeling assured that he will be able to prolong his life and mitigate human suffering, hesitate to sacrifice a few thousand or even millions of the lower creatures that he may gain the knowledge necessary to accomplish the desired end? Shall he allow a foolish sentiment to deter him from seeking in the animal kingdom for the facts which will enable him to confer a lasting benefit upon his fellow-men?

The revolting records of the various schools of philosophy and the still more horrible ones of the private vivisectors answer in the negative, and we would suppose from the work being done in the laboratories that the whole animal creation might be of little importance if only the welfare of man could be benefited, that the physical health of man was the *summum bonum* of existence. So man makes it his profession, his darling passion to search for the secrets of organic life and vital force under the covering of skin beneath which Nature has hidden them. Claude Bernard's ideal vivisector is

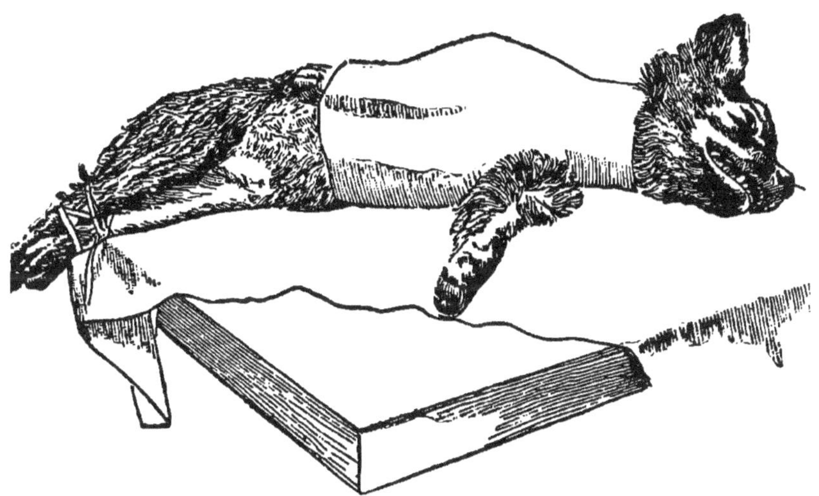

A New York doctor twisted and bound the legs of dogs in unnatural positions; forced the leg of one dog over its back, binding it, and sealing it in plaster-of-paris; kept it thus 145 days. The above illustration is an exact copy of the drawing accompanying the article written by said doctor and published in *Laboratory Researches*. They who know the pain of a limb even a short time in a cramped position can imagine the sufferings of this dog.

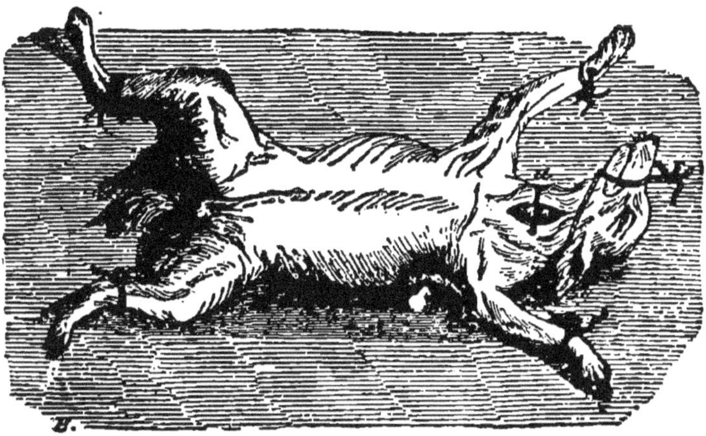

Windpipe of a dog dissected out to stop the cries of the animal under experimentation.—De Graaf, No. 5.

the man who "does not hear the animal's cries of pain, who is blind to the blood that flows, and who sees nothing but his idea and the organism which conceals from him the secrets he is resolved to discover."

A man thus dominated by one idea, with a sublime enthusiasm for the welfare of the human race, and believing in his heart that the lower animals were created solely for the use of mankind, and that the sacrifice of an unlimited number that one throb of pain may be saved to the human family, would have little compunction in submitting the poor creatures to the most exquisite torture in order that he might find one fact of value to the scientific world. Thoroughly absorbed in studying the law of life, he would consider that the means not only justified the end, but regarding it from a merely scientific standpoint he would believe that the great multitude of facts thus added to the store of human knowledge more than overbalances the amount of pain inflicted. To the devout physiologist the operation which to the uninitiated seems barbarous and horrible in the extreme is quite right and necessary.

The poor animal, bound upon the table, writhing in agony, whining piteously, lovingly licking the hands that inflict their wounds and begging in their poor dumb way for mercy, is a sight to make the basest, most hardened heart to throb with emotions of pity and regret—pity for their pain, regret that the faithful dog, the noble horse, the affectionate, purring cat, the animals which have been the aid, the companions and the pets of man should be subjected to such barbarities in order to gratify a monstrous and abnormal passion for scientific investigation which has fastened itself upon modern thought like a leech. Scientific men everywhere seem to have a contemptuous disregard for animal life, and to think that the poor dumb brutes were created solely for the purpose of furnishing them with material upon which to exercise their devilish ingenuity. Filled to overflowing with a curiosity which they will gratify at whatever cost, and utterly devoid of pity, they perform the most agonizing experiments, inflicting upon a sentient creature, whose nervous system is as highly organized as is that of man, the most exquisite torture. It is true they do not confine their work in this line upon animals entirely, for the pauper in the hospital often makes a good subject for an experiment of this kind, and medical men, while devoted to the interests of humanity in the abstract, care almost as little for individual suffering as for pain inflicted upon the poor dumb brutes. Putting aside any feeling of pity, sentiment or compassion for present distress, and caring only for the future generation which shall profit by his discoveries, he goes on his horrible work torturing to death countless innocents in order that he may add somewhat to the accumulation of facts written upon the pages of science.

Does even the highest good of man justify this wholesale murder—nay, worse, this damnable torturing to death of helpless brutes? Shall we so highly esteem the sufferings of man as to count as a light thing the horrible suffering among the lower animals that his pain may be mitigated in even the slightest degree? What is man more than his fellow-creatures that his comfort or even his life shall depend upon the facts gained at such a fearful cost? Is it not the basest cowardice, the most horrible cruelty that such a thought should dominate the mind of man? For his own physical well-being he is willing to sacrifice all the rest of the animal kingdom.

With utter disregard for pain in those he considers below him in the scale of creation he purchases knowledge at a fearful cost.

Ah, those poor science-tortured brutes! That they might but speak with man's tongue as spake the ass of old, and forbid the madness of men and teach them wisdom; that they might cry out with one voice against those scientific horrors which are classed under the name of vivisection, and that these physiological investigations which involve unutterable sufferings might be forever abolished!

Cardinal Manning once said, speaking of England's law for punishing cruelty to animals, "That law seems to express a great moral principle that people have no right to inflict pain." Surely it is true that we as superior animals have no moral right to subject to the most exquisite cruelty helpless beings, either for the purpose of gratifying our own scientific curiosity or saving ourselves from the effects of a gross and careless life. One can not but feel a passionate indignation toward one who, through blind ignorance and utter disregard for the simple hygienic laws which should govern his life, and of the moral and mental powers which should shape his destiny and make him but "little lower than the angels," brings upon himself disease, sickness and premature death, and then, instead of turning to the great physician for health and strength, seeks in this art of torture, this superior species of butchery, relief from the ills which make our life a curse. If he would but remember that through his own greed, his lust and his disobedience to the loving dictates of Nature he has fallen into sin, and would he be temperate in his methods of life, practice the simple laws of health as taught him by his Creator, he would not need to seek in the entrails of birds and beasts, in tendon and ligature, in brain and spinal-cord for the vital force, the elixir of life, the true essence of being, but would find all about him the principle which would excel in virtue the famous magic fount of the credulous Spaniard.

We have spoken in a previous chapter of the unmitigated horrors of the slaughter-pens, of cruelty practiced upon animals during shipping and confinement, in order to obtain animal foods for man, but no species of cruelty will compare with that which is practiced by scientific experimenters and the practicing physicians of this ultra cruel age. The number of poor creatures daily sacrificed to science by baking, boiling, freezing, by electricity and by the keen knife of the butcher is beyond belief. Claude Bernard's famous oven for ascertaining the exact amount of heat required to roast a dog would, perhaps, be a surprise to most people. Language would fail us if we attempted to describe the savage excesses of cruelty practiced upon creatures whose helplessness should have entitled them to pity. Some are roasted, some are stewed, some are sawn asunder, some are slowly picked to pieces bit by bit, the operator carefully following the lines of the different ligaments, nerves, arteries and veins, tracing them with as much care as if they represented vast stores of wealth. The sufferings of the poor dumb brutes can be better imagined than described. Not often are they put under the influence of an anæsthetic, and even when this is done, to again quote from the words of that world-famous butcher, Claude Bernard, "Curare is used. This is an anæsthetic which paralyzes the powers of motion, but leaves the power of suffering actually increased." Could aught of which the mind of man may conceive be more atrocious than this? A poor beast fastened securely upon a dissecting table, without the power to move but with every sense rendered doubly sensitive by the administration of a drug. Could anything be more horrible than this type of cruelty, which is the result of ignorance and a reckless disregard for the higher life of the community that such atrocities be perpetrated in the name of humanity? Shall this pandering to scientific curiosity be allowed? Surely some day the mind of man will revolt at the horrible practice and make an end of it forever. Surely some day society will emerge from this age of barbarism and will demand for the so-called lower creatures immunity from torture. When we think what the world might become were the tiger passions bred out, and man should stand once more in the divine image, we grow weary at heart, for we see plainly that it will be a long time before we reach that period. We can not refrain from wishing for the pen of a Milton or a Shakespere, the tongue of a Brooks or a Beecher, that we might tell to all people the horrors of the vivisection tables, the awful sufferings of those poor creatures doomed to so fearful a death as awaits them when once they fall into the hands of these fiends among men.

The tales of horror contained in the works of Claude Bernard, Paul Bret, Brown Sequard and Ricket in France, of Goltz in Germany, Mentegazza in Italy and Flint in America are sufficient to convince any one of the fiendish horrors that are committed in the divine name of Science. A mere glance at the tools employed by the vivisectors will be ample proof of the fearful lengths to which their trade has been carried, while the pictures of the unfortunate animals, securely bound and writhing in agony, is a shocking proof of man's barbarity. The fact that human misery may be mitigated to a slight extent by the facts gained from the practice of these outrages, will hardly compensate for the great amount of suffering undergone by the victims of man's cruelty.

The various societies for prevention of cruelty to animals have made many unsuccessful attempts to have vivisection abolished. The physicians, jealous of every infringement upon their "rights," while often denouncing the practices of these refined butchers have, nevertheless, opposed any enactment which will in any way interfere with their pre-established privileges. So they have stood with them, and upheld them in their horrible work in this "Coward Science," as Adams so aptly terms it.

If indeed vivisection had given us any valuable facts not obtainable from other sources, there might be some excuse for the reverence in which the "art" is held to-day by the vast majority, but as it has never added one fact to the truths of science, has not revealed anything that might not have been obtained in a manner less cruel and ghastly, all right-minded people must feel it a duty to denounce vivisection as neither scientific, useful, moral nor in any way justifiable. The animals operated upon are in an abnormal condition when placed upon the table, for the whole system must at once be affected. The sensations of fear and dread which must necessarily follow upon his being bound to the table, as well as by the evident preparations for some unusual occurrence, excites him to the highest pitch of terror. How can one expect to obtain a very satisfactory idea of the organic functions of a man by watching the effects of his torture of a poor, mangled, dying brute. True, man is fearfully and wonderfully made, but he is not made upon the same general plan as is the cat, the dog, the cow or the horse.

How then will aught be gained that is of value to the human family by thus slaying thousands of poor, wretched creatures? Vivisection has as yet done nothing for humanity, and surely nothing can be gained by this brutal method of interrogating Nature. In the very principle of the thing, man has no more right to perform painful or injurious experiments upon animals than he has upon human beings. Their sensitive organizations render them extremely susceptible to the slightest pain. The old, exploded theory that animals feel less pain than human beings has led many to make extremely severe experiments; but whether it is true that animals really have more or less sensations of pain than man, it is true that they suffer excruciating agony during the experiments.

We might be asked what is the effect of this life spent in the work of torture upon the vivisector himself? It is a well known fact that the scenes in the laboratory are of so shocking a character that the spectator has to exercise a strong control of his will to subdue the involuntary sense of horror which he experiences upon viewing his first operation. This soon passes off if the young man is determined to conquer his feelings of disgust. While it is true that vivisection does not demoralize a man in all ways, yet he must from the very nature of the case become so thoroughly heartless, so inhuman toward his kind and so brutal that he can never be again the tender-hearted, loving man he was before. The leaven of cruelty can not but thoroughly permeate his very being, and work upon his inward life and impress its stamp upon his face. That which engrosses a man's thought will find expression in his soul, and that which dominates his life will leave its imprint upon his whole being. This is one of the unchangeable laws of Nature.

The effect, then, upon his offspring, and the effect upon the race is one which demands attention. It is not for ourselves alone we live, but after us shall come another generation—let us hope a stronger, gentler, nobler one—whose aim in life will be the bettering of the conditions of all, whose watchword will be justice and whose ideal will be truth. Let us hope, too, that he will not be so selfish as to imagine that all things that are have been created solely for his benefit, and that he will accord to every sentient creature the rights he claims for himself. Truly he would question the right of any superior intelligence to experiment on him, that they might thereby gain information which would enable them to live more comfortably.

I find it a disagreeable necessity to speak upon the subject of woman's cruelty to the animal kingdom. While she does not directly cause the murder of innocent creatures, she does, by her demand for their furs and feathers as well as by her silence upon the subject, encourage this wholesale slaughter. She desires the luxurious garments which can be furnished by no other means, and seems to feel that she is justified in obtaining them at whatever expense, even while she shudders at the recital of the cruelties practiced upon the poor victims. Since the discovery of a method of converting the dull gray fur of the seal to a rich and lustrous brown the demand for seal skins has increased with wonderful

Rabbit paralyzed by hydrophobia inoculation in the Institute Pasteur. Exact copy of drawing from the Paris journal *L'Illustration*, 1891. Seen by Herbert J. Reid, of London, Feb. 22, 1894. Mr. Reid says, "The wretched animals were lying on their sides slowly dying of paralytic rabies, their hind legs extended and powreless but their eyes turned pleadingly toward the visitor."

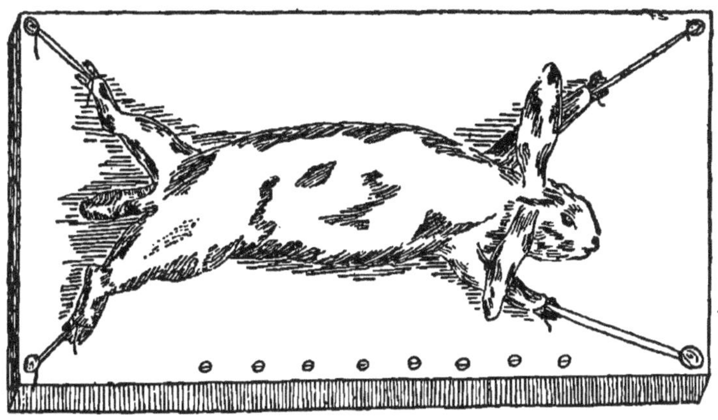

Rabbit bound for trepanning. Scene in the Institute Pasteur. No anæsthetics were given the animals.

rapidity. The shocking details of a seal hunt is enough to make one grow sick at heart. As these animals are more easily captured while cubs are young and helpless, the cruelties practiced upon the dams in dispatching them and procuring the skins is not the worst feature of the case—the little ones are left to perish with cold and hunger. An article commenting upon the hunting of seals, says: "The time chosen for the hunting is unfortunately the very period that of all others should be kept close. Except for a very short part of the year the seal lives to all intents and purposes on the open sea. But the female, when about to bring forth, seeks the shelter of the shore, where she suckles and watches her cubs until they are old enough to shift for themselves. At this time, wherever there are seals along the coast, herds of them will be found from a quarter to half a mile inland. The proportions are very much those of a drove of deer. The main body will consist of females, each with one or two helpless ones, while the males keep about the outskirts of the flock. As soon as a herd of this kind is spied, the boats are manned, and the whole vessel's crew, armed with bludgeons and axes, starts upon a 'cutting out expedition,' at the horrors of which humanity may well shudder. The only way to effectually kill a seal with completeness and dispatch is by a heavy blow with a bludgeon or a deep cut with an axe, so as either to crush or sever the nasal bones; and when the boat's crew have got ashore an indiscriminate slaughter is commenced, the whole herd being often butchered before a single one can reach the water's edge. The adult quarry is skinned with all possible haste, and as often as not with the life still in it. The cubs, who lie moaning and whining by the side of their dams, are knocked on the head if big enough to give their fur any value, and if too small to be worth the skinning are left without even the mercy of a *coup de grace.* Old seal hunters tell us, and we can well believe it, that it takes a man some time to get used to such cruel butchery, and that the half-human wailing of the little seals, as they climb and roll about the mangled carcasses of their mothers, is a sound that until he is hardened to the work will make a man's sleep uneasy at night."

There is really no excuse for this slaughter of animals and birds in order to obtain for women a more becoming dress. Everything she wears seems becoming, and the beautiful fabrics woven from the products of Nature render her much more charming than do the furs and plumage of tortured, sentient creatures. The woman decked in her priceless fur cape, with ornaments of rare birds upon her hat, represents much of suffering and cruelty. Still she seems deaf to the piteous cries of the poor struggling creatures which furnished the rich furs, and cares not that the sweet voice of the song bird has been hushed forever. Many are ignorant of the suffering which must fall to the lot of the poor victims thus sacrificed upon the altar of vanity. No man will admire a true woman more because she is decorated in the skins of wild beasts and has about her throat and upon her head trophies of the hunt. Only too often the skins of birds are torn off while yet alive in order to preserve the brilliancy and beauty of the plumage. All who have watched the habits of these dainty creatures, and have listened to their exquisite music, must feel a throb of pain when they see one of the tiny things wired into an unnatural attitude and perched over the fair brow of some delicate, refined woman. The Chicago *Tribune* of

November 28, 1897, printed an article which describes some of the shocking cruelties practiced upon the innocent victims of fashion, which we herewith insert for the benefit of those who know nothing of these horrible atrocities:

"Woman this fall has decked herself in a veritable garniture of death. Not to mince matters, it may be said plainly that the dress and the decoration of the woman of fashion just now is the product of cruelty which in some of its forms is almost unspeakable.

"Among the women who realize what the present trend of fashion means to the brute creation there have been formed Audubon societies for the protection of birds and for discouraging that fad of the millinery world which would place birds of song or plumage upon the brim or crown of every bonnet in this broad land. The attention of those women who have made the cause of the birds their own has been called within the week to the sufferings attendant upon procuring for feminine use other materials for dress and hat embellishment.

"For some reason other women who were apparently utterly untouched by the stories of the sufferings of the songsters have felt so sharp a pang at the recital of this new horror that with aigrettes streaming in the wind from their bonnets' tops, or with the plumage from some soft-breasted bird from the sea nestling just above their brows, they have started on a crusade against cruelty to one of earth's four-footed animals whose covering of wool has been declared by the fashion-makers as one of the things to be desired above all others for dress decoration purposes.

"It is the little Persian lamb, or rather the little Persian lamb's mother, whose story of suffering seems to have stung many women to the quick. Just who the native Persian naturalist may be who has shown a love for animals is not known, but undoubtedly American women will find him out, and before long there will be societies formed bearing his name and whose object will be the discouraging of the wearing of lamb's wool.

"It is barely possible that it may be hard to convince the ordinarily tender-hearted woman that she has about her the garniture of death, but it is not a bit hard to begin at the top and prove the case. There are more feather-crowned hats worn this fall than ever before, and woman has allowed her milliner to make her believe that the feathered kingdom suffers no loss, because the majority of bonnet decorations are made from the bodies of birds of prey.

"The truth is that these predatory creatures are the most interesting in the whole bird world, and it has been proved conclusively that they do just as much good to the agricultural world in the killing of harmful rodents and large insects as do the birds whose voices are lifted in song. The aigrette is more fashionable than ever. One hat in every three bears one of the white plumes of these innocents, whose killing is attended with suffering for the parent bird, to be followed by the slow starvation of the nestlings. The garniture of death starts at the crown of woman's bonnet, though, perhaps, the evidences of bird murder can hardly be said to represent the crowning cruelty.

"Woman has thrown out of her list of comparisons that which likens innocence to a lamb. With a bit of Persian lamb's wool as a covering for her bonnet, or with a pair of Persian lamb cuffs warming her wrists, or perhaps with a Persian lamb jacket 'snugged' to her body, she does not like to think of the innocence of the creature whose life was lost along with that of its mother that she, woman, the highest type of God's creatures, might be made beautiful. It is the fleece of the Persian baby lamb, the unborn baby lamb at that, which is just now the most fashionable article of woman's wear. Costly? Yes, so costly that only those who are rich may have a blouse of the wool made, but as every woman must needs have a bit of the fleece, if only enough for a collar or a bit of trimming, the demand is great enough to insure a continuation of the cruelty of the methods attached to procuring the wool for an indefinite period.

"The mother lambs are fed upon food which long experience has shown increases the delicacy of the fur of its unborn young. After a certain time the ewe is killed and the skin is taken from that which would have been its offspring, and this skin, the delicacy of whose covering would under other circumstances be a delight to any eye, is put on the market as an article to be sold to the mothers and daughters of the world. The skin of the lambkins has about the breadth of that of the rabbit. They are known to the trade as broad-tails. They are to be seen in the window of every shop where woman's needs or woman's luxurious tastes are the subjects for consideration.

"In New York the other day there met the members of a great society composed wholly of men whose object it is to prevent the wanton killing of birds for any reason whatever. Speaking of the habit that women have of adorning themselves with the bodies or skins of dead animals, the President of the society said that the practice arose from the fact that woman did not have strength of mind enough to be willing to act individually. No one of them, he declared, was willing to do anything by herself, but must needs wait until she could get a throng of her sisters to act with her.

"Mrs. John A. Logan, on the other hand, speaking of this very question of the wearing of the fleecy skins of the unborn Persian lambs, says that the use on the part of woman arises simply from ignorance of the methods used in obtaining the adornments.

"The facts about the case of the Persian lamb have become known to some little extent in Chicago, and inquiry among a number of dealers has shown that some women have refused point blank to buy the fleece, although acknowledging its beauty, the refusal being on the ground that the cruelty attendant on obtaining it was known.

"For dress trimming Persian lamb's wool might well be called the double garniture of death.

"There are scores of members of the Audubon societies from Boston to San Francisco who wear featherless bonnets on their heads but wrap their bodies in the skins of the mother seal and its baby. A prominent woman member of the Illinois Audubon society when playfully accused of hypocrisy by a bird-wearing sister because the Audubon member had on a sealskin sacque said in reply that seals did not sing in one's garden nor did they rear their young in the bushes under one's windows. Then she was chided again for not holding that cruelty at a distance was as wrong as cruelty practiced at home.

"There was, perhaps, some justice in the reproof from the woman who would wear birds, Audubon society teachings to the contrary notwithstanding. The demand of woman for the fur of the seal has led to international complications, and at one time actually threatened to bring about war; but even the prospects of bloodshed, coupled with the certainty of the cruelties in seal capture, did not and could not apparently abate one jot the desire of woman to wrap herself in the fur of these sea creatures, whose innocence is only less than that of the lamb. When the seals are upon the islands with their young they are attacked by men bearing spears and clubs. The seal is at a great disadvantage on land, and before the herd can reach the water the members are slain, often with the most needless cruelty. The creatures witness the slaughter of a few of their kind, and, knowing the fate which awaits them, they fairly beg for their lives.

"Sometimes the older seals are killed by a lance thrust into the heart, but the younger ones are invariably struck over the head with a club. The seal butchers do not stop to see whether the young are dead or not, only satisfying themselves that the blow has been sufficiently hard to prevent the stricken creature from making its way to the sea. Then the butchery goes on until every member of the herd has been struck or stabbed. Then the men go back leisurely and give the finishing touch to such of the seals as may have survived the first blow.

"There are humanitarians in America, who knowing the ultimate fate of the seals, wish that they might be exterminated at once rather than have the cruelties prolonged, and this even though the woman of the future might thereby be compelled to go without what now forms the greater part of the garniture of death."

It is hard to find a woman's heart 'neath the mass of

trumpery with which she has decked herself, but when once it is touched with a feeling of tenderness, her nature awakened to the suffering of any living creature, she will stop on no half-way ground, but demand a cessation of the cruelties and utterly refuse to in any way encourage that which she deems wrong. If all could but be impressed with the sense of injustice and could but realize the intense suffering caused by their foolish adoration of the idol, Fashion, this cruel slaughter would soon be abolished and the animals allowed to propagate their kind in peace, while the glad song of beautiful birds would echo and re-echo throughout the length and breadth of the land.

"Greed for gold" is perhaps the real foundation of this shameful trade in skins and feathers. It is not love for women that prompts men to brave the dangers of the Arctic regions and the torrid heat of the tropic zones to bring back the trophies for her pleasure. They serve a fairer mistress, and for her sake they will brave tempest and calm, cold and heat, fatigue, peril, death. The return voyage of the successful hunter is not gilded by dreams of fair women rendered more charming by the treasures with which he shall deck her person, but by visions of bank-notes and glittering gold which will be his when he has disposed of his cargo to the crafty tradesmen.

Neither do the fashion-makers and the tradesmen care whether their efforts make women beautiful or not providing they can but succeed in creating a demand for their wares and fill their pocket-books to overflowing with that which they value more than honor, more than life.

It is a terrible thing to think that men and women

HORSE FASTENED IN HOLDER.

This animal had his legs amputated at the knee joints for some "humane" (?) and surgical purpose no doubt. Being fastened in a holder shows beyond a doubt that he is alive and unanæsthetized. Anæsthetics are never used, being considered unnecessary as the animals are so tightly fastened as to be incapable of moving.

A live dog bound for experimentation.—From *La Physiologie Operatoire*. CLAUDE BERNARD.

will thus become dead to all the higher laws of their beings, and willingly slaughter thousands of animals to gratify a foolish vanity and an abnormal appetite. It is not because they obtain greater satisfaction from eating flesh, from wearing feathers and furs, or from sacrificing their fellow-creatures to the cause of science in order that they may ascertain facts regarding human anatomy than they would by living natural lives. Fashion dictates what shall be worn, and all bow to her decree. The Scientist, impelled by a morbid curiosity, begins to war against the lower creatures, and immediately has a following of enthusiasts. The fact that both are unnecessary and criminal carries no weight with the devotee to society or the savant.

But few think upon the subject, and many who do are indifferent to the suffering of either animals or human beings. Such things do not trouble them so long as they are comfortable and at ease with the world. They care not for that which injures others so long as they themselves escape from all inconvenience. Why should they feel a pang of regret for the pain of an Arctic seal or the agony of a brilliant song bird in the antipodes so long as they can command the money to gratify their tastes. Fashion dictates that they shall wear such ornaments; they simply obey.

It is not only the suffering inflicted upon sentient creatures, but the brutalizing of the human heart by the barbarities committed upon those who can not protect themselves from their tormentors that prompts these words. Man injures himself far more than he does the lower creatures when he thus turns murderer, and the effects of his actions upon the coming generation will be still more appalling than upon his own life,

It is not only that animal life is sacred, but the destiny of the human race is worthy of consideration. Shall man go on in the old, brutal ways, sacrificing not only the lives of his fellows, but brutalizing the nobler nature within himself, or shall he listen to the cry of the tortured ones and stop this ghastly outrage against nature? Words are inadequate to portray the horrors which abound everywhere. The pen of man is too weak to show in all their horrors the scenes of suffering about us, not only in the animal kingdom, but in the human family. Would that it might cease, and man, regarding as sacred the life of others, regain his own lost kingdom.

CHAPTER VII.

STUDIES IN FOODS.

I HAVE previously spoken of the effects of a vegetable diet upon savage animals kept in confinement, and noted how their whole temper might be changed by again allowing them meat. Is it then to be supposed that if animals are thus affected by the quality of food consumed, that the nature of man may not be changed in the same way? Fruits, vegetables and flowers are also affected in coloring and flavor by the chemical properties of the soil from which they spring. Horticulturists and floriculturists are enabled to produce fruits of exquisite flavor, and flowers of most brilliant or delicate coloring, as they may desire, by giving them the proper chemical nourishment. You have perhaps yourself noticed the wild flowers of the same species growing in different localities. Those upon the hill were perhaps delicate in hue, while those down near the stream were a rich, deep shade. You may have wondered why the fruit on one tree in your garden was a pale pink in color, while that grown upon another tree of the same stock, but planted in a little different soil and receiving a different quality of fertilizer, would blush rosy red, while the difference in flavor would be no less marked than that in coloring. So it has been with your berries, and the difference in the coloring of garden flowers are so great as to attract the attention of even the least observing. If fruits and flowers are thus affected by the chemical properties of the material from which they draw their life, will not the susceptible nature of man be affected by whatever enters into his physical organization? Considered from a physical standpoint alone, the theory is as we see uncontrovertible, and when we remember that man alone of all forms of life contains within himself a faculty which is most powerful in its effects upon his own outward nature, we can have some idea of the importance of the food he eats, and the manner of obtaining the same. If it be gleaned from the fields, plucked from the bending tree and swaying vine, if the simple juice of the ripened grape and the crystal water be his drink, he will hardly develop the ferocious traits which characterize the animals whose natural food is the prey for which they must lie in wait and capture by superior cunning and power. All instincts of the carnivora are ferocious, and have been developed through necessity. So it is possible for man to develop the cruel instincts of his nature, and to become as ferocious as are the most savage of the animal kingdom. Or it is possible for man by cultivating his higher nature, by subsisting on natural foods and by controlling his appetite, to grow more pure and noble. The possibilities of human development are practically boundless, and he may make life what he chooses.

Through ignorance of the laws which govern all things man has sinned against his own nature, and it is our hope that he may be brought to a knowledge of the truth, and that he may change his method of living. Men do not consider that they are themselves to blame for the deplorable state of affairs now existing, but by a little careful thought it will be found that intemperance lies at the foundation of all suffering. Self-gratification has reigned supreme in the hearts of the human family for ages. Deformity, imbecility and disease have been the portion of those individuals and those nations which have been governed alone by their appetites. The masses have ever failed to realize the real source of their many misfortunes, have regarded God as the author of their woes, and have bowed with what patience they could command under "divine dispensations of Providence," but have made little or no effort to restrain the hand of their tormentor, by obeying the dictates of Nature, and paying attention to sanitary conditions. The laws of our being are violated by crowding the stomach with unhealthy food, demanded by a morbid appetite, and the constitution must sooner or later become impaired. The human family has grown more and more lustful. Men eat and drink to excess, and inflame their passions by rich and exciting condiments, spirits, etc. They give themselves up to an abominable idolatry, and become diseased, corrupt and ferocious. Like the Israelites of old, they have grown to loathe the "light food" furnished by an invisible force, and long for the flesh-pots of Egypt, preferring to live under the galling yoke of a hateful bondage rather than practice temperance. They are well punished for their gluttony, for following close upon their feasting come disease and death. Even while the meat is between their teeth they are stricken down. Generations past appetite has controlled reason to an alarming extent. Intemperance in eating and drinking, and the indulgence of the baser passions have benumbed the nobler faculties, for it is not the physical health alone which is injured by excessive indulgence in eating and drinking, but the mind is affected, and the finer sensibilities are blunted. The material of which the body is built must of necessity be pure if it is desired that the mental faculties be in a perfect state. Grossness in foods leads to grossness in manners, grossness of manners to grossness of thought. If animal food, reeking with filth, is the material chosen, is it not reasonable to suppose that the body will be unclean? Watch the swine in his filthy pen, or wandering about the fields, a dirty scavenger, and then after you have seen him killed and dressed and made ready for the table, decide whether you would wish your body to be built of the material he has gathered, or to reflect in your life the psychic force which will be generated by eating his flesh and making it a part of your life.

Here and there, in every highway and byway of the world, in every village and hamlet, and in every condition of society existing among flesh-eating nations, is to be found the terrifying monster we call disease. Disease is born in sin. It therefore thrives and subsists in iniquity. It is an unnatural condition. It is something which should be so rare as to mark every individual who has permitted it to fasten itself upon him as a social, moral and physical leper. Health is the natural condition of every living creature. Animals, be they high or low in the scale of organism, in their natural condition live out their terms of life and die natural deaths. Few human beings indeed live a natural life and die a natural death. A very large proportion of the human race commit suicide with their teeth by eating unnatural food and drinking unnatural drink. They live unnatural lives, generate unnatural offspring and die unnatural deaths.

Natural lives can be lived only by understanding and obeying the laws of nature, by eating man's natural food, the grains, roots, fruits and nuts ; by drinking only man's natural drink, pure water ; by breathing pure air by day and by night, by observing habits of personal cleanliness, by taking sufficient exercise to promote digestion and excretion, and by avoiding worry and fretting. The man who will live such a life as this will enjoy health in its greatest perfection, and will have before him the prospect of a long and happy life. Inheritance may interfere with the perfect development of many, for the sins of their parents may be visited upon them even to the third and fourth generations, but notwithstanding this, all may attain a large measure of the blessings resulting from natural living, and those who are not handicapped by hereditary infirmities can attain to such exalted conditions of life as but few indeed have dreamed as being even remotely possible.

The disease by which the middle and latter part of the lives of more than one half of the human race is embittered is due to avoidable errors in diet. Children are fed upon meat almost from infancy. In their early childhood they are permitted the use of coffee, tea, rich pastry, confections and a hundred and one digestion retarding and fever exciting foods and drinks, which the genius of modern times prides itself upon having invented.

Modern cookery is poisoning the race, not to the extent of immediate fatalities, but to a degree which restricts mental development and induces sickness and disease in forms and manners of manifestation which are innumerable. The human misery for which it is responsible is beyond calculation. The sick, the half sick and the slightly indisposed constitute by far the larger portion of living men and women of all nations of the earth.

Robbery, rapine, murder, vulgarity, debasement, perversions of the intelligence of the body are traceable almost entirely to the gratification of morbid appetites, which have grown from perversions in the matter of diet.

The instances of people who have feasted and died from the effects of inordinate eating, are innumerable. Observers of mortuary records can bear ample testimony to the fact. How often do we see recorded in the newspapers the death by apoplexy of some prominent man, who dined the evening before with friends, and after a hearty dinner and convivial drinking, fell prostrate before he could leave the scene of feasting; or who, after dining, retired to sleep with a stomach distended, with blood sluggish, with painfully laboring heart, his entire system clogged with gluttony, and who, after thus taxing nature beyond the possibility of further endurance, has fallen into the sleep from which there is no awakening. Let us pause and consider. Let us think of the gains we will make by correcting our habits of living. If we have been taught as children to indulge in excesses of improper living, let us exert the strength necessary to deliver ourselves from the damning effects of custom and the mistaken ideas we have been taught to accept as natural and matter-of-course.

It is largely in our power to induce or banish sickness, to live happily or miserably, to lengthen or shorten our lives, and the food we eat is the principal factor to be considered. As soon as we begin to live, we begin to die, and would die if we were not sustained by nourishing food. Waste of organic matter in the physical system is caused by every word, thought, breath or action, and tends to the ultimate destruction of the whole body. We should eat food to repair this waste, not to glut and

EDWARD MARTIN.

(See page 204.)

gorge the system so as to hinder the natural processes of elimination, restoration and recuperation.

We are continually wasting away, and consequently require to be renewed and rebuilt, otherwise we would die in a very short time. Now comes the important question as to what food is best suited to repair the waste, nourish the body and promote health and long life. It has been determined by experiments and scientific observation extending over a period of many years, that the grains, nuts, fruits and legumes constitute man's natural food. Your physician may not tell you this. Your physician may be a man who conforms in his medical practice to the modes of society, who, having familiarized himself with human anatomy and *materia medicæ*, is prepared to write an arbitrary prescription for drugs as a curative where certain conditions are indicated in accordance with arbitrary rules for observation and symptom-reading, without any reference to the food that shall sustain the body in order to make possible the curative processes desired. If he is such a man, and lamb chops are fashionable, he will recommend that you eat lamb chops. The probabilities are that he will say nothing about diet unless asked. If asked, the chances are that he is incompetent to recommend a diet, from the fact that he has given it but very little special study. Therefore it is quite useless to apply to medical authority for correct instruction as to what you should eat or drink, because each physician will order or forbid that which is the fashion to order or forbid. The seeker after long life and happiness must of necessity be his own judge as to the quantity and quality of food best suited to his own particular case, but the source from which the selections must be made, if we wish to grow healthy and retain the use of the faculties to an advanced age, is the vegetable kingdom.

There are many reasons why a diet consisting solely of fruits, nuts, roots and legumes is the best for mankind. The whole internal and external structure of man clearly indicates his natural unfitness to live on flesh, while on the contrary, by the use of fruits, nuts, grains, etc., as food, the processes of digestion and assimilation are carried on in a most natural manner, with less liability to derangement, and with the greatest possible nutritive qualities for the formation of bones, muscles and pure blood, and the most sturdy and vigorous constitutional development.

It is undoubtedly a fact, which may be stated without any fear of successful controversy, that under the most favorable circumstances the flesh of animals can never be freed from impurities. The processes of waste and repair are constantly going on in the living system, and there is, therefore, always present in the tissues and blood-vessels more or less broken-down cell-structure which is on its way to the excretory outlet of the system. If these processes are arrested by the death of the animal, this effete matter remains in the flesh, together with that large proportion of the impure venous blood which is in the capillaries, and hence the flesh is more or less loaded with impurity. This would be true under the most favorable conditions—if the animal had not been overfed, and if it for a period preceding had been at complete rest for a considerable time before its slaughter. It is more particularly true of animals which have been subjected to the strain of long-distance shipping, or to the exertion and excitement of driving to the slaughter-pen. The strain and excitement cause in these animals a more rapid consumption of tissue, which, as it is consumed, generates effete and poisonous matter in large quantities, which will necessarily be arrested in the process of elimination at killing, and will be retained in the fibre of the meat when it is placed upon the dining-table for consumption. Furthermore, a large proportion of the animals slaughtered for the public market are diseased. Lumpy-jaw, Tuberculosis and various other diseases of cattle have, in recent years, been very common, and while various Board's of Health and other means have been created for the purpose of preventing the slaughter of diseased animals, there is absolutely no doubt that thousands upon thousands of animals in various stages of disease have been slaughtered and consumed.

A few years ago a butcher died as a result of handling the meat of an ox taken dead from the stock-car. The fact that the butcher died from mere contact with the flesh of this diseased animal did not prevent it being sent to the market and sold as food. It found its way into many unsuspecting mouths, and, for aught we know, many deaths may have been caused from its consumption. Such might have been true with but the slightest possibility of the real cause being traced, as the habit of flesh-eating has become so fixed upon the people that very few of them ever think upon the subject, and many who do so are unwilling to consider it as a possible vehicle of fatality.

Improper feeding is another prolific cause of disease among animals. It can hardly be reasonably assumed that the slop-fed cattle of the distilleries would be good food even if, under favorable conditions, animal flesh was a proper diet for man; yet thousands upon thousands of slop-fed cattle are marketed every year. The unvarying testimony of those who refrain from eating flesh is that they enjoy apparent immunity from disease. Many after adapting a natural diet of fruits, nuts, grains and legumes have been relieved of long-standing complaints, such as Gout, Rheumatism, Indigestion, Paralysis and other dangerous diseases, while others have been able to enjoy a life of comfort by the relief which this diet affords. Persons practicing this diet quickly recover from accidents and surgical operations, and it is a fact that has long been demonstrated that a non-flesh diet will cure old-standing ulcers, tumors and dropsy where all other means have failed. Excessive drinking and shameless debauchery are almost completely unknown among people who make a practice of subsisting upon a purely vegetable diet.

Physical conditions show that the blood of flesh-eaters is loaded with impurities in a state of decomposition, which result in Rheumatism, Gout, Boils, Tumors and many aggravating ailments of the skin, liver, kidneys and intestines. So far as exhaustive investigation has disclosed, those adhering to a natural diet have never been known to suffer from Scrofula, Salt Rheum and other skin affections.

It is impossible to inculcate into the minds of children a love of animal nature, and the fear of giving pain to any living creature so long as they are inured to scenes of slaughter and bloodshed, and are taught that the slaughter and sacrifice of lambs, calves and fowls, with the suffering inseparable therefrom is both necessary and right. Power, endurance, courage and unusual capacity for toil belong to those non-flesh-eating animals which alone since the world began have been associated with the fortunes and conquests and the achievements of man. All the labor of the world is performed by the herbivora—horses, oxen, mules, elephants and camels. The carnivora have almost no place in the economic affairs of man, and have neither lightened his labor nor contributed to his progress. The carnivora do not exhibit the endurance of the horse, which toils with hardly any rest from early morning until late at night. The arguments in favor of the non-flesh diet are objective of greater human strength, greater powers of endurance, more perfect physical and mental development and higher spiritual attainment.

The life of man can be prolonged and his health and happiness insured by a diet of which the flesh of animals forms no part. Therefore, neither justice nor benevolence can sanction the revolting cruelties that are daily practiced in order to pamper perverted and unnatural appetites. Many of the greatest personages of the world, whose labors have resulted in making new epochs in the various periods of history, whose works have been uplifting, civilizing, refining and spiritualizing, have subsisted upon a purely vegetable diet. Among these are mentioned Buddha, Pythagoras, Plato, Epicurus, Seneca, Plutarch, St. James, Clemens of Alexandria, Porphyry, Chrysostom, the Emperor Julian and Prudentius, and while in later times we have Gassendi, Mandeville, Cocchi, Rousseau, Voltaire, Lord Chesterfield, John Wesley, Howard, Newton, Milton, Haller, Locke, Swedenborg, Ritson, Franklin, St. Pierre, Dr. Priessnitz, Dr. Lambe, Dr. Hufelund, Shelley, Byron, Phillips, Brotherton, Alcott, Graham, Lamartine, Michelet, Struve, Dr. Clubb, Dr. O'Leary, Dr. Stockaam and scores of others.

Aside from the vast array of demonstrable facts which can be adduced to prove the great benefits of a pure and natural diet, there is a religious aspect to this subject. The Bible says, in Genesis 1, 29: "And God said, behold, I have given you every herb-bearing fruit which is upon the face of the whole earth, and every tree in which is the fruit of a tree yielding seed; to you it shall be for meat." Here is a command in the very earliest divine history which says that the herbs and fruits of the trees shall be for meat. The command is plain and unequivocal. The fact that flesh is not mentioned makes it presumable that there was no contemplation on the part of the Creator that man should become so perverted in the purity of his original nature as to fall into the error and sin of flesh-eating. Another Bible passage recites that "every living thing that liveth shall be meat for you even as the green herb, for I give you all things; but flesh with the life thereof, which is the blood thereof, shall ye not eat."

How can the modern Christian, with any claim to consistency, make his daintiest delicacies and luxuries of flesh food, and how can he, particularly, relish the flesh of filthy swine, which authorities agree is expressly forbidden by the Bible? Many of the most learned scientists of the age declare that man is a frugivorous or fruit-eating animal, and that he does not possess either teeth suitable for tearing or digestive apparatus naturally adapted to its assimilation. The teeth and digestive organs of the carnivora are especially fitted for the functions required of flesh-eating animals. Upon these statements it must be assumed that the voluntary consumption of bodies of dead animals in civilized countries is a violation of the fundamental laws governing man's existence, and therefore totally unnecessary. People of observing dispositions, who have had opportunities to study, and there are few of us who have not, can to the last one testify to the cruelty and inhumanity that is widely exercised in the slaughtering-houses and pens. Such incidents as that of kicking cows, twisting their tails and maltreating them until their eyes burst before they can be gotten into the slaughter-house are not infrequent, and it is of very frequent occurrence that animals have been deprived of food and water for from two to ten days before they were driven to slaughter. If a man have a God and worship him, how can he look with acquiescence upon such scenes?

If the W. C. T. U. and all the temperance organizations and workers in the country could realize the potency of the curse of flesh-eating as an inducement of that other curse, alcoholism, the system of temperance work would be revolutionized within a fortnight, and every earnest worker in the cause would become an active advocate of proper diet reform. Abstinence from flesh is of far greater importance than abstinence from strong drink, for the former includes the latter and goes far beyond it in the work of reformation. It may be stated as an almost unfailing result that drinkers who can be prevailed upon to abstain entirely from the use of flesh, will, in a little time, become abstainers from choice. It is stated that not a single drunkard can be found among non-flesh-eaters.

Unimportant as the subject may seem to many, yet the fact remains that the health, prosperity and morality of individuals and nations depends almost altogether upon the food and drink of those individuals and nations. There is more nourishment in a piece of bread, providing it is properly prepared, than in meat. There is more real life in a glass of water than in a glass of whiskey, more value in the fruits and nuts than in all the dainty concoctions of the baker. There is more material for blood and bone and tissue in simple wheaten cakes than in the choicest dainties of the *chef.*

In choosing your foods you have an abundance from which to select. You can take the vegetables fresh from Mother Earth, full of the life-giving principles; the fruits, dewy with the rains from heaven, foodfit for the gods; plain, wholesome bread and the nectar which Nature gives, or you can choose your food from the slaughter-pens. You will find dead animals of every description exposed for sale, from the patient, worn-out horse to the dwellers of the sea.

Behind the scenes in the butcher-shop and the slaughter-house the sights are even more revolting than those in the market-places. If people could only believe the truth of what is told them regarding the horrible things sold there the custom of flesh-eating would be abandoned.

The flesh of horses is now quite a common article of diet, and much of the "beef" sold in the markets is horse-flesh. Almost all the old, worn-out dray-horses are utilized in this way. When no longer able to work they are sent to the slaughter-house, and a glance at their poor, pitiful bodies as they are led into the pens is enough to make any one who witnesses the sight a vegetarian for life. Many of the cattle are in the same miserable condition, but they must be killed soon, else they will die naturally, so they are hurried to the market.

The horrible diseases which too often afflict these poor creatures are, of course, one of the sources of the many

DAVID H. REEDER, M. D.

Author of "Home Health Club" Books.

maladies which afflict the human race. The meats which are inspected and pronounced unfit for use, simply for form, are then hurried into the markets to be disposed of before they become quite putrid. Can we wonder at the loathsome and incurable diseases which defy all medical skill when we see the filthy things that are used for food?

It is shocking, but as true as it is shocking, that cats, dogs, rats and other animals which have so long been pronounced unclean, are daily sold in our markets under the guise of game. Many are ignorant of the fact, and dine in blissful unconsciousness of the fact that they are eating the things which are supposed to form part of the diet of heathen and savages.

There are some men in every large city who make it their business to go about picking up dead poultry and selling it on the streets. In a city everything is utilized by someone. The garbage boxes are made to yield their treasures to the eager gleaner, and everything is turned to account. Many of these people who have gathered dead poultry from the refuse claim to be direct from the country, and thus often obtain a handsome price for their wares. Of course this is a disgraceful state of affairs, but what can be expected of those who are so steeped in sensuality that they are dead to all the higher things of life?

It is no wonder that people resort to mustards and sauces to give flavor to such noisome messes as are placed before them for their daily consumption; no wonder they drink quantities of liquor to allay the burning thirst which such unnatural foods produce; nor is it any wonder that they become gross in thought and habit after living on such diet for years.

It is to the credit of most people who live upon animal foods that they have never given this subject of foods much thought, for if they did but study the subject, they would see the error of the way they are now pursuing and live differently. It is not greatly to their credit, however, that they give no thought to those subjects which so closely concern their own welfare and that of the whole human family.

If every one knew the true history of every animal he helps to eat, I feel confident but little flesh would be consumed; if every one could study for himself the barbarities of the shipping-pens, the slaughter-houses, the packing-houses and the markets these relics of savagery would soon be abolished. Few indeed have ever studied the subject of diet, but it is safe to say that all who have studied it intelligently have been convinced of the necessity of a pure diet for the promotion of health and morality among the people.

Some who read these pages will doubtless exclaim: "I do not believe it!" But, alas, it is too true. They have but to investigate to assure themselves that the half has never been told, simply because it is impossible for language to depict the horrible scenes and describe the fearful results of this hateful and degrading custom of slaying and eating countless innocent ones whose lives are as precious as our own.

From these scenes of carnage, from this hideous spectacle of death, we will turn to the peaceful pursuits of man that furnish him with the food which Nature intended for his use. We see him engaged in tilling the soil, in planting the grain, in caring for the crop as it comes to perfection; then we see him gather into the garner the products of his labor. The animals are his companions and fellow helpers. With their superior strength they assist him in accomplishing that which would otherwise be impossible, and he, with his superior intelligence, cares for their comfort. The thought of converting them into food does not come into his mind. He is content with the simple fruits of his toil. The waving grain, the golden fruits, the rich nuts, give him greater pleasure than would the flesh of his co-laborers, and his natural methods of living bring him peace and health.

His glorious intellect is not clouded by the effects of gluttony, nor the delicate tissues burned as with raging fever by alcohol. He is in full possession of every faculty, full of strength, hope and courage. No longer a slave to his appetites, he commands his powers and is able to accomplish what would otherwise be impossible. He has no need of the physician's advice, for Nature teaches him the laws of health; no need of patent medicines and poisonous drugs, for he is obedient to the divine dictates of a higher law, and lives in harmony with her teachings. He has no need of a mediator between himself and Nature, for she disdains not to teach the humblest of her children, and to lead them in the right paths. He has no fear of old age, for he has lived temperately and knows it will not be painful. He does not dread death, for he knows that it is but the outworkings of natural law which can not be changed.

Living thus in accord with his higher nature, he scatters light and love in his pathway, beautifying and glorifying life until it is divine, filling the future of the race not with tears and pain and despair, but with health and love and happiness.

Is it not true that man born of woman has a right to be born of healthy parentage, has a right to a heritage of noble instincts, good moral nature, a lovely disposition, a healthy body and a strong intellect? Has he not a right to demand of his creators purity, and shall he be denied that which is his by divine law? Man's duty is not to himself alone. It is not enough that your own life be above reproach. You have a higher duty—a duty to others. You dare not shrink from it.

While in ignorance of the law man is absolved from conscious guilt of sin, though he can not escape the penalty of the law; but when it struths are presented to him in such a manner that he can not possibly misunderstand them, he is no longer guiltless of crime. We forgive the unlearned, saying pityingly, "They know no better;" but there can be no excuse for those who knowingly violate their own nature and perpetuate in their offspring the diseases and vices which result from their hateful habits.

It is sadly true that but few realize the true nobility of life, or recognize the fact that man makes his own life and the lives of future generations what he will. Few indeed study the workings of the subtile forces which he sets in motion, whose influence he can never control, whose power he can never wholly realize. He knows nothing of the wonderful processes which, working quietly within the soul of man, molds and shapes his thoughts; nothing of the delicate forces which deep in his being are making his life grand and true, or base and ignoble; nor does he know that everything which comes into his life leaves its indelible imprint upon the soul.

There is but one way in which humanity can rise above its present environments, its present limitations, and that is by universal education along all lines of human economy. Man sins against himself through ignorance of his true nature, his grand possibilities and his noble destiny. So long as he remains ignorant, so long must he of necessity remain degraded. So long as his soul sleeps in sensuality, so long must humanity be steeped in viciousness and crime. So long as the fathers and mothers of the race are responsive only to the lower vibrations of life, so long must their offspring partake of the nature of the parent tree. So long as lust reigns supreme, so long will the higher forces of man be held in bondage by the things that delight the senses alone.

It is a long way, would you say, from the food of man, from his daily habits, from his occupation and surroundings, to the higher spiritual life which should glow within his soul? Nay, the relations are very close, very delicate, and yet clearly perceived by the careful student. Man is a wonderful creature, endowed with possibilities that are as limitless as is the universe, and nothing but ignorance of his true nature prevents him from claiming his own. Knowledge can come only by obedience to the laws of his being, and those laws he does not obey.

A great change is undoubtedly coming, for throughout the world the fact is being impressed upon the minds of thousands, the conviction that life is a sacred thing not to be sacrificed except upon the ground of necessity; that cruelty is crime; that the human body, created to be the temple of God, should not be polluted by being converted into a sepulchre for the dead bodies of slaughtered beasts, and that the true Christian spirit is one of universal benevolence, which demands the sacrifice of self in place of the sacrifice of others.

Philanthropists, religious leaders, humanitarians, temperance reformers, and all persons who desire to lessen the world's misery and increase the sum of human happiness, are urged to consider earnestly what should be their attitude toward the great reform which is foreshadowed in the natural movement, and to possess themselves of the indisputable facts which are known with relation to the benefits of a radical dietary reform. The general acceptance of these ideas will solve most of our social problems and usher in the most beneficent revolution that the world has yet seen. They will ultimately prevail, because they are true, and every man and woman to whom they are presented must individually decide whether to hinder or to hasten this change of thought and custom which promotes and exerts a stupendous influence upon our own and coming generations. Nature has been sadly perverted for many centuries, and we trust that the time is near when you may resolve, and, resolving, inaugurate for you and yours a reform which shall be to the lasting benefit of all the generations that are yet to come.

Apparatus for studying the "Mechanism of Death by Heat."—Bernard's *Lecons sur la chaleur Animale* (Paris) p, 347. Living dogs, rabbits and pigeons were thus baked or boiled to death while their tormentors make notes of their suffering and death.

Vivisectors deceive and quiet the public by claiming these horrible acts are "for the benefit of mankind" and in "the interests of science," but there is nothing learned that in the remotest way could be of use in the treatment of human beings.

CHAPTER VIII.

EFFECTS OF DIFFERENT FOODS.

THE food of man is of vital importance. The body is built up of the material taken into the system, and reflects exactly in texture and coloring of the skin, in the appearance of hair, nails, teeth, and, in fact, in every portion of the physical being, the quality and quantity of the food introduced into the system.

To many this may seem an exaggerated statement, but to the earnest student of life it is as easily explained as any other fact in Nature. The laws which govern the process of converting a given material into flesh, blood, bone, tissue, thought and psychic forces are the same which govern all other processes of life. As effects always follow causes, and as the effect must of necessity be to a certain extent like the cause which produces it, it is easy to trace the relation between the food we eat and the outward physical nature.

If man's food is coarse and gross in quality, and if his habits of eating and drinking be as gross as is his food, he can not but become coarse and brutal. His figure grows corpulent, his flesh is dull, heavy and puffy, his eyes are blood-shot, and his whole aspect betokens the glutton. He is heavy and swinish in his movements, the expression of his countenance is sleepy, and his brain is slow in action. You can distinguish him among a hundred as a man devoted to his flesh-pots, his bottle and his pipe.

Contrast him with the man of abstemious habits. Note the clear skin, the delicate changing color, the quick, keen eye, the comprehensive glance, the freedom of movement, the strong spirituality, the life, vigor and power of the one as compared with the stupidity of the other. Note also the difference in the mentality—the one is awake to every issue of life, the other is satisfied when his comfort is undisturbed. One is confident of himself, and knows well his own power of accomplishment, the other neither recognizes his ability, nor cares to exercise whatever talent he may possess; so long as he feels not the pangs of hunger, he cares little for the inward man; so long as his dinner, his pipe and his mug are ready to his hand, he does not trouble about the future.

The question presented to your mind would be, "What makes the difference in the life and thought of these two men?" Upon careful examination of their modes of life, you could not but decide that the food eaten by them was one of the causes of difference in thought and appearance. In this you would be quite right, for while one deluged his stomach with all manner of gross material, the other subsisted upon the simplest foods, partook of them sparingly, and in consequence was physically and mentally clean. While one flooded his system with poisonous drinks, the other drank only of the nectar provided by Nature.

There are many works now before the public setting forth the relative values of different articles of food, and it is hardly necessary to treat the subject in an exhaustive manner in this book, but I wish to touch upon a few of the most salient points. My chief object in all the books I have written is not to lay down rules to be blindly followed, but to stimulate individual thought, to encourage people to get out of the habit of accepting, without question, statements made by those whom they consider competent to judge for them, and to think, weigh and decide for themselves. If after due deliberation you find any statement to be true, incorporate it into your own life; but if you allow every man to be your judge, and rely implicitly upon the doctrine advocated by the last writer you have read, you will never be able to know exactly where you stand, nor will you ever be able to believe anything for more than a short time. You should, however, give a subject careful study, and never condemn what you do not understand or what you have not proven.

You can well afford to be called a "crank," to endure gracefully the sneers and jests of your companions and friends, to even be subjected to some personal discomfort in order to live what you know to be the higher life, and to develop the powers of soul with which man alone is endowed. To become spiritually pure it is necessary that you be personally clean; to be personally clean and physically perfect, it is necessary that you build up your system from only the best material obtainable. I have shown in another chapter how the thoughts affect the body, and I wish now to assure my readers that the foods taken into the system exert upon the thoughts an influence quite as great as that of thought upon the outward man. Beside the effect of food upon thought is the influence exerted upon the physical being. There is a direct affinity between the nourishment and the physical nature. This is a fact which has been recognized by all the teachers of old.

Can a person live without meat? is often asked in a

most incredulous tone and manner. We reply at once that people not only can live without meat, but enjoy the very highest state of health. This ultra position we are perfectly aware is in direct opposition to the popular belief, and also most strange is the fact, to the teachings of the medical profession, which we should naturally suppose, in the view of the light of science, observation and experience of the present day, would be the pioneers of a doctrine of a correct life. A great majority still most tenaciously cling to the greasy and swillish flesh-pots of Egypt, and contend to the last for the old habits of life, which any man with brains can see is everywhere making society a practical hospital.

Can people live without meat? We repeat they can, and find a degree of health never experienced under the grossness of an animal regimen. Facts illustrating the truth of this statement are in existence in every part of the world, and are repeated every day.

Who does not know that more than half the inhabitants of this globe seldom taste of animal food from the cradle to the grave. Yet that such is a fact no person of intelligence dares risk his reputation in denying. And the nations who thus live are notoriously and proverbially healthy and long-lived. They occupy, too, every variety of climate, and almost every geographical position on the earth. None can escape these facts.

Go back into the history of nations, and what is the condition of things? We shall find our position fortified by an amount of evidence entirely overwhelming; that people, wherever their locality, and in whatever point of time they have lived, have been famous for superiority in every respect whose diet has been simply natural. This holds with the philosopher, poet, statesman, in fact, through every grade of the given people.

In our own age and time, what nations are the heartiest, strongest and longest lived? Go to the barren heights of Russia, and see what a physically enduring race of men and women are there! See the almost incredible number over a century old, and in the enjoyment of vigor, comeliness and general integrity of powers which are never the boast of our people after forty! And this almost entirely from a coarse bread and water diet—never any meat. Ye gourmands of the flesh-pots, can ye equal these? Never!

Look at the Hungarians, a nation the prowess of which is the song of the world. See their hardihood, their prodigies of valor, their marches, their heroic sufferings. Where is the people who can surpass their bodily fatigue? And yet, ninety-nine in every hundred of the Hungarian army seldom eat an ounce of meat. Courage, patriotism, fortitude, strength like theirs binds its soul in a pint per day of oat-meal and water.

How with the Irish? Where under the blue and spangled firmament shall we find a more robust and physically vigorous race? And who does not know that the great body of the nation live mainly on a scanty diet of potatoes?

It is the same with the Polish and Peruvian peasantry. More hardy races or people, who can endure more extended fatigue, who are more active, cheerful, kind, can not be found. Do they eat meat, roast-beef, pork, swim in soups and bathe in grease? Parched corn is almost their entire food. Poor benighted ones, they know no better use for the beasts of their glorious mountain-sides than honest service till death.

But we might go to the circle of the nations, and find

similar facts to meet us. Wherever the fair and ruddy banner is unfurled, there, and there only, we see this same simple, satisfactory, common-sense style of life. And we do not see it where other systems prevail. England, France, America, and parts of some other nations, are the principal homes and nurseries of disease and death.

In our own country, wherever natural diet has been adopted, it has been with the most gratifying results. We need look for no other issue when the experiment has a fair and timely trial. The belief in its superior advantages is happily gaining ground every day. It seems to us that nothing can sooner induce a defection from the flesh-eating ranks than the demonstrated fact that most of the meat sold throughout our markets and from our carts is actually diseased. This every person may know with but little exertion. The very idea is perfectly abhorrent and sickening that people will persist in devouring from day to day that which cannot be other than full of corruption and death. And yet, it is so. Why, it is a well known fact that there are thousands round all our cities and large towns whose sole business it is to buy up old hulks and living carcasses of animals which have been worked to within an inch of death, and then fatten them by a most cruel system of stall-feeding, till the animals themselves are ready to take their last foul breath when natural death is saved by the hurried axe and knife. And this is, to no little extent, the kind of flesh that runs its purifying blood through our markets, and over which the bloated face of the gourmand chuckles with such sensual delight.

Not only the physical, but the intellectual system would thrive with far more purpose and energy under a natural diet. Whatever insures elasticity, energy, vig[illegible] to the body has a faithful reflex on the mind. This position is practically acknowledged by all men of mind, for where anything of an intellectual task is performed, the assistance of a coarse, spare fruit and vegetable diet is put under arbitrary contribution.

The horrible and filthy process of "stuffing" animals and fowls before sending them to market is one of the chief sources of diseased meat. You can readily imagine the fearful condition of the bodies of these animals after they have thus been fattened, by drawing upon your imagination and thinking of the condition your own body would be in if you were yourself shut in a pen or coop, allowed no exercise, surrounded by various kinds of foods, and given drinks which would create a desire for an inordinate amount of eating, and besides what your appetite called for, were regularly stuffed until it was impossible for you to hold another mouthful, almost impossible for you to breathe. Think of the days and weeks of this treatment. The whole system would become overburdened and diseased, the blood would become sluggish, the various organs would be unable to perform their proper functions. The waste-matter is not discharged but is thrown back upon the system to be reabsorbed and to either cause disease or to create superfluous fat. You would become dull, stupid, burdensome to yourself, a great unwieldy thing which you would not recognize as your true self. Every organ of the body would feel overburdened. Would you call this a state of health, or would you feel that your whole being had been outraged?

This is the condition of many of the animals which find their way to our markets. It is quite impossible that they be healthy when such conditions prevail, and it is equally impossible for man to partake of their diseased flesh and retain his health.

In this age of wild speculation and reckless theorizing, an appeal to facts is exceedingly serviceable in illustrating the more consistent views of the dietetic reformer. The clamor of the votary of animal indulgence is loud and almost appalling, assailing all changes in diet as dangerous innovations. We are often assured that a disuse of animal food will inevitably destroy our vital energies and result in imbecility. We are again told that, though the man of sedentary habits may survive for a time, vegetable and farinaceous diet will not sustain life and afford the necessary stamina for the more laborious avocations of life. How far this is true must be decided by the verdict of enlightened observers, those who have had experience in both systems, for such only are prepared to judge understandingly.

Many people have tried a natural diet, composed of only those things which nature herself furnishes from her bounteous stores, have lived in accordance with her behests, and have found it not only possible to exist upon such a diet, but that their whole nature was purified and ennobled, while their mental powers were increased many times. Men of culture find their minds growing stronger, and they are able to work to much greater advantage; laboring men find this diet fully satisfies every demand made upon them. They grow stronger, more wholesome, and can work with much less fatigue than when much of the energy of the body was necessarily devoted to the work of disposing of the superfluous material, and also in a vain endeavor to

Yours Truly
F. E. Ormsby

(See page 221.)

undo many of the injuries inflicted upon the system by improper diet and improper methods of eating and drinking.

There are a great many theories abroad in the land regarding the question of foods, diet, etc. There are many conflicting statements made by learned men on both sides of the question. It is for every one to study the subject thoroughly, and to decide for himself what is best. After careful study and wide experience, I have formulated a method of natural diet which I know will meet the requirements of all. What has proven beneficial to the greatest number for a long time past must of necessity be best for the whole human race. "Vegetarianism" is very good, and is a step in the right direction, but abstinence from meat alone is the end at which they aim, and, while an important point, is not the only one to be considered. In my system, which I believe to be as nearly perfect as possible, I have not forgotten the spiritual and mental life while dealing with the physical. It is almost impossible to treat upon one subject without indirectly treating of the other two; therefore, when the vegetarians advocate abolishing the horrible and filthy butcher-shops I am in hearty sympathy with them, for the diet advocated by them will do much to decrease disease, filth, crime and drunkenness, and every impulse toward the better life is so much in favor of humanity as a whole.

It is an interesting study to watch man as he gathers and prepares his food. He sows and reaps and gathers into gárners, he hauls his wheat and corn to mill and brings with him the fine white flour. He fattens his pigs and his oxen for slaughter, and pickles and dries and salts away the meats, renders out the grease, and with sage, spice and other relishes, packs away the daintiest parts for use. He stores his vegetables and his fruits, his nuts and berries. He has herds and flocks to furnish him with butter, cheese and milk, while numerous fowls give him an abundance of eggs, beside often serving to satisfy his appetite. He obtains vast amounts of fuel with which to render palatable the stores thus gathered, and builds great ovens for their preparation. He hires a bevy of cooks and waiters to concoct messes and to serve him. Then such a sputtering, boiling, baking, stewing, roasting and mixing! Such combinations of ingredients, such an amount of trouble to fill the stomach and ruin the health! No meal is perfect without meat. Huge joints and sides of slain are served at every table. Meat of itself, reared, slain and prepared in the manner before described, is sufficient to give any one dyspepsia, to say nothing of other diseases which it may produce, but when spread with mustard, saturated with sauce or strong gravies, it is a wonder to me that any man can eat it and retain even a semblance of health. Add to half-cooked meats, rich soups and heavy gravies, the inevitable white loaf, the cakes, pies, puddings, creams, butter, milk, cheese and various vegetables saturated with grease, and washed down with coffee, tea, chocolate, ice-water, and strong liquors, wine, beer and whiskey, and you will begin to be appalled at the amount of labor which we demand of our digestive organs. It seems almost incredible that their powers of endurance may be taxed so greatly and yet so quickly recover and take up their work again.

After the stomach has been abused for years, and the whole system is thoroughly depleted by the treatment it has received, and begins to give warning that it can stand no more, man rushes off to a doctor for advice. He is met with a grave face and a wise shake of the head, is given a prescription, pays his fee, takes his medicine, gets worse, for he still goes on in the old gluttonous way, eating and drinking as much as before, never dreaming that it is his gross habits of life which have caused his trouble. Added to the work previously demanded of his digestive organs is the disposition of a daily potion of poison, which they must get rid of as best they may. After trying various prescriptions and obtaining no relief, the glutton next turns his attention to the various patent medicines which flood the market. Tries them all and gets worse. He is now a miserable, morbid creature without hope, and while he finds it impossible to eat as much as formerly, he still disposes of as large a quantity as possible, and it is generally of such a quality as to render his existence miserable in the extreme.

These people never give their mode of living credit for having brought them into this abject state, but regard it as a "special dispensation of Providence," and many of them suffer uncomplainingly, believing that "whom the Lord loveth He chasteneth," and expecting a particularly bright crown and an exalted position very near the Throne of Grace as a compensation for the years of misery with which they have been afflicted by divine ordinance. That their ill-health is due to their gluttony and carelessness they could not be persuaded. Neither will they be persuaded that they can cure themselves—that they can grow strong and well by observing the simple laws of hygiene and living in accordance with the dictates of Nature. They consider ill-health a sort of punishment sent upon them, and never once consider that they have brought it upon themselves by their folly and disobedience.

Many people would live simple, healthful lives did they but understand the laws which govern them, but as they have never been taught, they go on in the way of their parents. Others would not conform to any rule however beneficial they might know it to be. They lack will-power and self-control, and would not make an effort to guide their appetites even if they were fully persuaded they could thereby recover perfect health and strength. To them the present effort is too difficult and they will trust to the future to make them well or they will bear their lot in silence.

If man would return to the primitive way of living, would exercise properly in the open air, would dress comfortably and in conformance with the requirements of health, would cease to worry, would sleep and rest more, would live a simple, natural life, subsist upon the products of the soil, and would practice self-control in all things, a new era would dawn upon the human race. There would be no more need of the many hospitals, prisons, houses of correction and work houses, for poverty, which is the mother of crime, could not exist among a people who were temperate in all things. Crime could not flourish in a community where purity of soul was universal and disease would find no existence among those who were physically clean both within and without. If the soil were used to raise grains and fruits for the sustenance of the people instead of being turned into huge pastures for raising stock and broad fields of billowy grain for fattening them for the market, thousands now out of employment would turn their attention to agricultural pursuits and earn sufficient bread for themselves and their families. Drunkenness would cease to exist, for men would no longer inflame their appetites with stimulating foods which create a craving for strong drink.

With a change in diet, we can see almost limitless possibilities for the human race. Freed from the chains of poverty, from the curse of vice, from the degrading influence of liquors, from the murderous tendencies which have for so long dominated humanity, and from the thousand and one ills that have kept the race down in the depths of weakness and woe, man would be delivered from the curse he has brought upon himself by yielding to his intemperate desires. The old Bible story of the doom pronounced upon the first pair because of their gratifying their appetite has indeed been fulfilled. Man has ever indulged his desires without thought of future consequence, he has disobeyed the first commandment of his nature "thou shalt not partake of the fruit" and by his sin has multiplied sorrows, even unto the children of the third and fourth generation. He has ignorantly given himself up to the gratification of those things which were pleasing to the eye and exceedingly to be desired, knowing that in so doing he was violating the highest dictates of his being, and has recked not of the future but thought only of the present. Down through the ages has echoed that accusing voice, majestic and sorrowful, "Because thou hast done this thing, cursed art thou," and mankind has groaned under the doom pronounced, although teachers, sages and philosophers have earnestly endeavored to point out the way of truth and to teach him to walk therein with all simplicity.

Never until man recognizes his own innate greatness and realizes the true relationship existing between himself and the universe, and knows that in abusing his physical being he also demoralizes his moral and spiritual nature, and by so doing degrades not only himself but also the whole race, will he make an effort to conquer self, to gain control of the forces which surround him and shape his destiny in accordance with his highest ideal of truth and love.

Purity of thought, word and deed, purity of body and soul should be his watchword. It is not so difficult of accomplishment as many suppose. It requires but to know self, to conform the life to the highest dictates of the individual conscience, to be temperate in all things, honest in all things, sincere in all things. The life must be clean, inwardly and outwardly. A pure soul can not dwell in a filthy body; so, if the physical nature be pure, healthful, strong, nourished upon the most wholesome foods, the moral and spiritual life will partake of its beauty. The unconscious influence of such a life upon others is very great. Perhaps no one fully realizes the importance of man's influence upon others. Even the personal appearance of an individual has its effect upon all who come within the radius of the surrounding aura. Instinctively you recognize a pure, healthy man, and your soul goes out to him in greeting. You wish to be more like him. In the same subtle way you know a coarse, brutal nature, an unclean physical and moral being, and shrink from the spiritual contact as you would from a vile thing. Your innate sense has warned you, and you immediately surround yourself with invisible barriers which keep at a distance all who are disagreeable.

If, as we have shown, the body is built up of the foods taken into the system, and is like it in principle, how necessary that we furnish it with none but the best materials, and care for it as though it were a thing of value. Plain, nourishing foods, carefully prepared, dainty, rich and clean, eaten only under favorable circumstances, are absolutely necessary to health, strength, beauty of mind and body. You can not abuse the body and expect it to make you a good servant, you can not afford to let it become your master.

This is certainly an important point to be observed by those who are laboring so earnestly and so hopelessly in the cause of temperance, and one they should not overlook. While hundreds cry for bread, thousands of bushels of grain which might be converted into rich and palatable foods are wasted by being made into vile liquor, which crazes the brain and destroys the self-respect, brutalizes and degrades the nature of the man who uses it. While thousands of children grow up in ignorance and vice, hungry, cold, dirty and miserable, robbed of every right to which innocent childhood is heir, men manufacture into poison the gifts of Nature to her children.

Surely there is something radically wrong where humanity is thus cheated of its right to life, liberty, health and honor. I feel no hesitancy in saying that if man would return to his simple habits of living, would eat only those foods which Nature meant that he should eat, and live in strict accordance with her laws, he would soon regain his lost "Eden" of health, temperance and purity.

The effects of alcohol upon the human system are well known by all who have given the subject even the slightest thought. You have but to walk abroad, and watch those whom you meet to perceive the deadly work it is doing among the victims of the wine cup. You have but to go into the alleys and byways of any of our great cities to learn of the fearful effect it produces upon human life. Besotted men, miserable women, filthy children swarm these vile places, and disease and death are everywhere.

If people but knew that these conditions are induced by the foods and drinks they take into their stomachs they would certainly be more careful of their diet. Ah, that my words might touch the hearts of all those who are thus making of life a hideous mockery, that my voice might reach to the ends of the earth and teach all peoples the lessons of health and strength, of beauty and love.

Many people believe that alcohol is strengthening, that it is beneficial if not partaken of in too large quantities. In this erroneous opinion they are, alas, too often confirmed by the physician who gravely advises them to "take a little wine for their stomach's sake" and their often infirmities, but only that those same infirmities may be increased and the possibility of their becoming patients for life assured.

It is true that alcohol taken into the stomach will produce a feeling of warmth and a general feeling of exhilaration. The brain is stimulated to increased activity for a time, but as soon as the effects have passed away the system is left in a state of collapse and the brain-power is weakened, while if too much is taken the brain refuses to work at all and the sensibilities of the man are deadened. The very fact that alcohol produces such a terrible appetite, and one that must be satisfied at whatever cost, proves it to be a poison of no ordinary stamp. That material which might be made into nourishing food for man is thus converted into that which kills his finer nature proves that we have not yet learned the most simple lessons in human economy.

How different from all the intoxicating beverages is that safe, simple drink prepared for man by loving Nature. Who has not observed the extreme satisfaction which children derive from quenching their thirst with pure water? Childen drink because their system demands fluid, and they are refreshed by the same law of life which created the desire for drink. Grown people drink whether they are thirsty or not, because they have found a way to make drink pleasant.

Strong meats, indigestible breads, which have been improperly compounded, greasy, ill-cooked vegetables, pies, cakes and all sorts of condiments create a craving for strong drink in large quantities. The more one eats and drinks, the more he wants to eat and drink. The heavier, coarser and grosser the material taken into the system, the stronger will be the craving for gross foods and vile drinks. The purer the food, the less it requires to nourish the physical man, and the less one becomes a slave to his appetite.

For the full and perfect development of the whole nature of man it is necessary that his food be not only pure in quality, but just sufficient in quantity to nourish properly every organism. The bones, nerves, brain and flesh each need the exact amount of food which will keep them in good working order. They do not need too much, for then the system becomes clogged and throws too much work upon some one organism. The human body is like a machine: if kept in good running order, properly fired, oiled and watched it does beautiful work; but as soon as the machinery, from neglect, from accumulation of dirt and grime on any part, becomes unfit for service, it can no longer be trusted. So with the human organism—so long as the fires of life are kept burning brightly and the body is kept clean, the functions of the body in perfect order, there is no cause to fear; but when the system is gorged with useless waste, and the appetites for strong drink and improper foods are constantly calling for satisfaction of their unnatural desire, how can body, brain or muscle be in a condition to accomplish its work?

Many people will say: "Well, I suppose it is not good for one to use liquor or tobacco, to drink coffee or to drink tea, but it has never hurt me, and, really, I can not do without it." They know it is not good for them to indulge their abnormally developed appetites, and yet they have not strength of will to combat them. They prefer to gratify their present desires rather than keep themselves clean, and pure, and healthful throughout life. They prefer the present to the future, and when the future comes they regret the past.

If every one could be induced to give up the use of liquors, tea, coffee, tobacco, meat and miserable messes called pastry, to live cleanly, temperately, righteously for one year, what a change would take place in the morals and manners of the world.

However, at the foundation of all unnatural conditions of life, of all intemperance, all drunkenness, all vice, all sickness and disease, lies that most horrible and unnatural thing—the slaying and eating of our fellow-creatures. From it comes everything that causes woe to the human race.

Every great philosopher, thinker and teacher of men since the first dawn of civilization has endeavored to show the evils of becoming a slave to casual appetites, and has proven conclusively that the debauchery of the human race was due to the disgusting and filthy habit of flesh-eating. All who have made a study of human life from its beginning up to the present day have the same story to tell. Man is debased by self-indulgence and his flesh-eating propensities have been the cause of all other forms of intemperance in which he has indulged.

This may seem a strange doctrine to many, but you have only to use your reasoning faculties to find that it is all true. You have but to watch the effects of different foods upon others, whether individuals or nations, to find the truth of all I say; to study the law of cause and effect to be convinced in your own mind that the desire for flesh has been cultivated in man, and has led to the desire for strong drink, has created disease, and has made human life a miserable farce.

Even the most hostile critics have been compelled to concede the sufficiency of grain and fruit to sustain the physical man, and have been forced to allow the effect of a meat diet in producing drunkenness and disease. They have been led to see that Nature always punishes the violation of her laws, and that for man's criminal disregard for animal life and his inordinate gluttony he has offended her dictates, and must suffer her displeasure.

It is only too true that with the increase in the ranks of physicians, and the multiplicity of medicines, diseases and deformities in the human race are rapidly increasing, and when we consider that the human structure is built out of the material selected, cooked and taken as food, we know the cause of all this sin and suffering. What is the remedy? There is but one, and that is proper food. As man is regenerative, he has the power to rebuild himself into natural conditions if he will but direct the forces at his command in an intelligent manner.

Man must be either natural or unnatural. If he live in an unnatural manner he must suffer from an inharmonious relation to the laws of his being, and one of the first conditions of being in harmony is that he select his own proper foods from those things provided by Nature for the sustenance of her creatures, prepare them in a rational manner and partake of just enough to keep his body in the most perfect condition. Only by so doing will the human race attain to a perfect state.

Dog suffering from inoculated rabies. M. Pasteur remarked to Mr. Mayet, "He will die tomorrow." Pasteur kicked the bars of the cage and the animal dashed at him with bleeding jaws, then turned and tore the litter of its kennel, uttering a piercing and plaintive cry.—*L'Illustration, 1891.*

"Pasteur is dead, thank God!"—MRS. C. L. H. WALLACE.

"Large dog on which various experiments have been made over night. It seems no worse."—*La Pression Barometrique*, p 637. PAUL BERT, Paris.

CHAPTER IX.

OPINIONS OF THE ANCIENTS.

MODERN writers are in the habit of quoting from the works of their illustrious predecessors in order to give force and authority to their own position. I am no exception to the rule, and desire to now call the attention of my readers to the writings of those whose names are household words, whose wisdom has made them the pride of their own nation and the admiration of other countries, whose goodness of soul and purity of life have endeared them to the hearts of all nations, and whose sincere endeavor to uplift and help their fellow-men has made their names loved and honored by all.

The fallacy that the conquering and dominant races have always been flesh-eaters has long since been exploded, and we can point to the Indian tribes of this country who, while a race of meat-eaters, yet went down like a field of wheat before the reaper when the superior skill and strength of a more civilized race were opposed to their animal cunning and cruelty. The ancient Greeks and Romans and ancient Egypt were long the recognized conquerors of the world, and their simple diet, their pure lives and their superior mental qualities were no doubt the reason they were so successful in commerce, so invincible in battle and so superbly grand in statesmanship. To be a Roman was to be a hero; to be a Greek was to be physically, mentally and morally clean;

to be an Egyptian was to claim sonship with the gods. When Greece and Rome gave themselves up to luxurious gorging and riotous drunkenness, their downfall was swift. The once proud mistress of the world lost in the revel of the wine-cup her honor and her greatness. Rome is deserted, Athens is in ruins and the curse of Nature rests upon all who disobey her, and brings to ashes the grandeur of empires.

Empedocles, one of the greatest of Latin poets, in his philosophical poem on Nature thus sings of the "Golden Age:"

"Then every animal was tame and familiar with men—both mammals and birds; and mutual love prevailed. The trees flourished with perpetual leaves and fruits, and ample crops adorned their boughs throughout the year. Nor had these happy people any War-god, nor had they any mad violence for their divinity. Nor was their monarch Zeus or Kronos or Poseidon, but Queen Kypris (the divinity of love). Her favor they besought with fragrant essences, and censers of pure myrrh, and frankincense, and with golden honey. The altars did not reek with the blood of oxen."

The name of Plato, the greatest philosopher the world has, perhaps, ever seen, deserves a conspicuous place in the history of Dietetic reformation. The Hellenic people in general were noted for their abstemiousness. Plato observes in his writings that, "The springs of human conduct depend principally upon diet," and was himself a most frugal man, living a pure and simple life. If the claims of being a philosopher must be based upon the following: first, an eager desire for all real existence; second, hatred of falsehood, and devoted love of truth; third, contempt for the pleasures of the body;

fourth, indifference to money; fifth, high-mindedness and liberality; sixth, justice and gentleness; seventh, a quick apprehension and a good memory; eighth, a musical, regular and harmonious disposition, then Plato was indeed one in the truest sense of the word. His life and thought were as pure as his food, his mental power and his physical being were sustained only by the natural products of a genial southern clime. The great and powerful influence of platonic thought upon succeeding generations need not be dwelt upon in this connection. It shows, however, the influence exerted by a pure life upon his fellow-men.

Plutarch was an ardent admirer of Plato and his school, but attached himself to no sect or system. From an Essay on Flesh-eating we quote the following:

"You ask me upon what grounds Pythagoras abstained from feeding on the flesh of animals. I, for my part, wonder of what sort of feeling, mind or reason that man was possessed who was the first to pollute his mouth with gore, and to allow his lips to touch the flesh of a murdered being, who spread his table with the mangled forms of dead bodies and claimed as daily food and dainty dishes what but now were beings endowed with movement, with perception and with voice?

"How could his eyes endure the murderous spectacle of the flayed and dismembered limbs? How could his sense of smell endure the horrid effluvium? How, I ask, was his taste not sickened by contact with festering wounds, with the pollution of corrupted blood and juices? 'The very hides begin to creep, and the flesh, both roast and raw, groaned on the spits, and the slaughtered oxen were endowed, as it might seem, with human voice.' This is poetic fiction, but the actual feast of ordinary life is, of a truth, a veritable portent—that a human being should hunger after the flesh of oxen actually bellowing before him, and teach upon what parts one should feast, and lay down elaborate rules about joints and roastings and dishings up. The first man who set the example of this savagery is the person to arraign; not, assuredly, that great mind, which, in a later age, determined to put a stop to such horrors.

"For the wretches who first applied to flesh-eating may justly be alleged in excuse their utter resourcelessness and destitution; inasmuch as it was not to indulge in lawless desires, or amidst the superfluities of necessaries, for the pleasure of wanton indulgence in unnatural luxuries that they (the primeval peoples) betook themselves to carnivorous habits.

"If they could now assume consciousness and speeoh they might exclaim, 'O blest and God-loved men who live at this day! What a happy age in the world's history has fallen to your lot, you who plant and reap an inheritance of all good things which grow for you in ungrudging abundance! What rich harvests do you not gather in! What wealth from the plains! What innocent pleasures is it not in your power to reap from the rich vegetation surrounding you on all sides! You may indulge in luxurious food without staining your hands with innocent blood, while as for us wretches, our lot was cast in an age of the world, the most savage and frightful conceivable. We were plunged into the midst of an all-prevailing and fatal want of the commonest necessaries of life from the period of the earth's first genesis, while yet the gross atmosphere of the globe hid the cheerful heaven from view, while the stars were yet wrapped in a dense and gloomy mist of fiery vapors, tempests and hail-storms prevailing, and the sun (earth) itself had no firm and regular course. Our globe was then a savage and uncultivated wilderness, perpetually overwhelmed with the floods of the disorderly rivers, abounding in shapeless and impenetrable morasses and forest. What wonder, then, if contrary to nature we had recourse to the flesh of living beings, when all our other means of subsistence consisted in wild corn (or a sort of grass) and the bark of trees and even slimy mud, and when we deemed ourselves fortunate to find some chance wild root or herb. When we tasted an acorn or beechnut we danced with grateful joy around the tree, hailing it as our bounteous mother and nurse.' Such was the gala-feast of those primeval days, while in other respects the whole earth was one universal scene of passion and violence.

"But what struggle for existence, or what goading madness has incited *you* to imbrue your hands in blood, you, who have, we repeat, a superabundance of all the necessaries of existence? Why do you belie the Earth as though she was unable to feed and nourish you? Why do you despite to the humanizing (divinity) Ceres, and dishonor the sweet and mellow gifts of Bacchus, as though you received not a sufficiency from them? Does it not shame you to mingle murder and blood with their beneficent fruit? Other carnivora you call savage and ferocious—lions and tigers and serpents—while you pollute your hands with blood and come behind them in no species of barbarity. And yet for them murder is the only means of subsistence, whereas to you it is a superfluous luxury. For, in point of fact, it is not lions and wolves we kill to eat as we might do in self-defense, on the contrary we leave them unmolested, and yet the innocent and the domesticated and helpless and defenceless, these we hunt and kill, whom Nature seems to have brought into existence for their beauty and gracefulness.

"Nothing puts us out of countenance, not the charming beauty of their form, not the plaintive sweetness of their voice or cry, not their mental intelligence, not the purity of their diet, not the superiority of understanding. For the sake of a part of their flesh only, we deprive them of the glorious light of the sun—of the life of which they were born. The plaintive cries they utter as they flee trembling in all directions we affect to take to be inarticulate and meaningless; whereas, in fact, they are entreaties and supplications and prayers addressed to us by each, which say: 'It is not the satisfaction of your real necessities that we deprecate, but the wanton indulgence of your appetites. Slay to satisfy actual hunger—if you must—but do not deprive us of sweet life merely to indulge your palate.' Alas for man's savage inhumanity! It is a terrible thing to see the table of rich men decked out by those layers-out of corpses, the butchers and cooks; a still more terrible sight is the same after the feast, for the wasted relics are even more than consumption. These victims then have given up their lives uselessly. At other times, from mere niggardliness, the host will grudge to distribute his dishes, and yet he grudged not to deprive innocent beings of their existence.

"Well, I have intercepted the excuse of those who allege that they have the authority and sanction of Nature; for that man is not by Nature carnivorous is proved in the first place by the external frame of his body—seeing that to none of the animals designed for living on flesh has the human body any resemblance. He has no curved beak, no sharp talons and claws, no pointed teeth, no intense power of stomach or heat of blood which might help him to masticate and digest gross and tough flesh-substance. On the contrary, by the smoothness of his teeth, the small capacity of his mouth, the softness of his tongue, and the sluggishness of his digestive apparatus, Nature sternly forbids him to feed on flesh.

"If, in spite of all this, you still affirm that you were intended by Nature for such a diet, then, in the first place, yourself kill what you wish to eat—but do it your real self, with your own natural weapons, without the use of butcher-knife, or axe, or club. No; as the wolves and lions and bears themselves slay all they devour, so in like manner do you kill the cow or ox with a grip of your jaws, or the pig with your teeth, or a hare or a lamb by falling upon and rending them on the spot. If, however, you wait until the intelligent existence be deprived of life, and if it would disgust you to have to rend out the heart and shed the life-blood of your victim, why, I ask, in the very face of Nature, and in despite of her, do you so feed? But more than this—not even after your victims have been killed will you eat them just as they are from the slaughter-house. You boil, roast and altogether metamorphose them by fire and condiments. You entirely alter and disguise the murdered animal by the use of ten thousand sweet herbs and spices, that your natural taste may be deceived and be prepared to take the unnatural food. A proper and witty rebuke was that of the Spartan who bought a fish and gave it to his cook to dress. When the latter asked for butter, and olive oil, and vinegar, he replied: 'Why, if I had all these things, I should not have bought it!'

"To such a degree do we make luxuries of bloodshed that we call flesh a 'delicacy,' and forthwith require delicate sauces for this same flesh-meat, and mix together oil and wine and honey and pickle and vinegar, with all the spices of Syria and Arabia, for all the world as though we were embalming a human corpse. After all these heterogeneous matters have been mixed and dissolved, and in a manner corrupted, it is for the stomach, forsooth, to assimilate them—if it can. And though this may be for the time accomplished, the natural sequence is a variety of diseases produced by imperfect digestion and repletion.

"Flesh-eating is not unnatural to our physical constitution only. The mind and intellect are made gross by gorging and repletion; for flesh-meat and wine may possibly tend to give robustness to the body, but they give only feebleness to the mind."

What horrors are here depicted with the touch of a master hand. Who after reading these words could again pollute his lips with the touch of flesh of murdered brutes? Who again would imbrue his hands with the blood of his fellow-creatures? Who can even glance at the poor senseless corpses without a shudder of disgust and a thrill of pity?

Seneca, the great moralist, regarded with peculiar horror the habit of flesh-eating, and did not hesitate to speak and write against it whenever opportunity presented. In writing to a friend, he says:

"December is the month when the city (Rome) most especially gives itself up to riotous living. Free license is allowed to the public luxury. Every place resounds with the gigantic preparations for eating and gorging, just as if," he adds, "the whole year were not a sort of Saturnalia."

He contrasts with all this enormous waste and gluttony the simple living of Epicurus, who, in a letter to his friend Polyænus, declares that his own food does not cost him sixpence a day, while his friend Metrodorus, who had not advanced so far in frugality, expended the whole of that small sum:

"Do you ask if that can supply due nourishment? Yes; and pleasure too. Not, indeed, that fleeting and superficial pleasure which needs to be perpetually recruited, but a solid and substantial one. Bread and pearl barley (polenta) certainly is not luxurious feeding, but it is no little advantage to be able to receive pleasure from a simple diet of which no change of fortune can deprive one. . . . Nature demands bread and water only; no one is poor in regard to those necessaries."

"How long shall we weary heaven with petitions for superfluous luxuries, as though we had not at hand wherewithal to feed ourselves? How long shall we fill our plains with huge cities? How long shall the people slave for us unnecessarily? How long shall countless numbers of ships from every sea bring us provisions for the consumption of a single month? An ox is satisfied with the pasture of an acre or two. One wood suffices for several elephants. Man alone supports himself by the pillage of the whole earth and sea. What! Has Nature indeed given us so insatiable a stomach, while she has given us so insignificant bodies? No: it is not the hunger of our stomachs, but insatiable covetousness (ambitio) which costs so much. The slaves of the belly (as says Sallust) are to be counted in the number of the lower animals, not of men. Nay, not of them, but rather of the dead. . . . You might inscribe on their doors, 'These have anticipated death.'"

The extreme difficulty of abstinence is generally alleged:

"It is disagreeable, you say, to abstain from the pleasures of the customary diet. Such abstinence is, I grant, difficult at first. But in course of time the desire for that diet will begin to languish; the incentives to our unnatural wants failing, the stomach, at first rebellious, will after a time feel an aversion for what formerly it eagerly coveted. The desire dies of itself and it is no severe loss to be without those things that you have ceased to long for. Add to this that there is no disease, no pain which is not certainly intermitted or relieved, or cured altogether. Moreover, it is possible for you to be on your guard against a threatened return of the disease, and to oppose remedies if it comes upon you."

On the occasion of a shipwreck, when his fellow-passengers found themselves forced to live upon the scantiest fare, he takes the opportunity to point out how extravagantly superfluous must be the ordinary living of the richer part of a nation:

"How easily we can dispense with these superfluities, which, when necessity takes them from us, we do not feel the want of. . . . Whenever I happen to be in the company of richly-living people I can not prevent a blush of shame, because I see evident proof that the principles which I approve and commend have as yet no sure and firm faith placed in them. . . . A warning voice needs to be published abroad in opposition to the prevailing opinion of the human race: 'You are out of your

PROF. URIEL BUCHANAN.

(See page 220.)

senses (insanitis); you are wandering from the path of right; you are lost in stupid admiration for superfluous luxuries; you value no one thing for its proper worth.'"—(Ep. lxxxvii.)

Again: "If the human race would but listen to the voice of reason, it would recognize that (fashionable) cooks are as superfluous as soldiers. Wisdom engages in all useful things, is favorable to peace, and summons the whole human species to concord."—(Ep. xc.)

"In the simpler things there was no need of so large a supernumerary force of medical men, nor of so many surgical instruments, nor of so many boses of drugs. Health was simple for a simple reason. Many dishes have induced many diseases. Note how vast a quantity of lives one stomach absorbs—devastator of land and sea. No wonder that with so discordant diet disease is ever varying. . . . Count the cooks: you will no longer wonder at the innumerable number of human maladies."—(Ep. xcv.)

We must be content to extract only one more of Seneca's exhortations to reform in diet:

"You think it a great matter that you can bring yourself to live without all the apparatus of fashionable dishes; that you do not desire wild boars of a thousand pounds weight, or the tongues of rare birds, and other portents of a luxury which now despises whole carcasses, and chooses only certain parts of each victim. I shall admire you then only when you scorn not plain bread, when you have persuaded yourself that herbs exist not for other animals only, but for man also, if you shall recognize that vegetables are sufficient food for the stomach into which we now stuff valuable lives, as though it were to keep them forever. For what matters it what it receives, since it will soon lose all that it has devoured? The apparatus of dishes, containing the spoils of sea and land, gives you pleasure, you say. . . . The splendor of all this, heightened by art, gives you pleasure. Ah! those very things so solicitously sought for and served up so variously, no sooner have they entered the belly than one and the same foulness shall take possession of them all. Would you condemn the pleasures of the table? Consider their final destination (exitum specta)."

If Seneca makes dietetics of the first importance, at the same time he by no means neglects other departments of ethics, which, for the most part, ultimately depend upon that fundamental reformation; and he is equally excellent on them all.

From writings of Chrysostom we quote the following passages showing the sensitive horror with which this most eloquent and estimable of the "Fathers" regarded the pernicious habit of flesh-eating. Denouncing the grossness of the ordinary mode of living, he eloquently descants on its evil results, physical as well as mental:

"A man who lives in pleasure (i. e., in selfish luxury) is dead while he lives, for he lives only to his belly. In his other senses he lives not. He sees not what he ought to see; he hears not what he ought to hear; he speaks not what he ought to speak. Look not at the superficial countenance, but examine the interior and you will see it full of deep dejection. If it were possible to bring the soul into view, and to behold it with our bodily eyes, that of the luxurious would seem depressed, mournful, miserable, and wasted with leanness, for the more the body grows sleek and gross the more lean and weakly is the soul. The more the one is pampered, the more is the other hampered. As when the pupil of the eye has the external envelope too thick, it can not put forth the power of vision and look out, because the light is excluded by the dense covering, and darkness ensues; so when the body is constantly full-fed the soul must be invested with grossness. The dead, say you, corrupt and rot and a foul, pestilential humor distills from them. So in her who lives in pleasure may be seen rheums, and phlegm, and catarrh, hiccough, vomiting, eructations and the like, which, as too unseemly, I forbear to name, for such is the despotism of luxury, it makes us endure things which we do not think proper even to mention. . . .

" 'She that lives in pleasure is dead while she lives.' Hear this, ye women, who pass your time in revels and intemperance, and who neglect the poor, pining and perishing with hunger, whilst you are destroying yourselves with continual luxury. Thus you are the cause of two deaths—of those who are dying of want, and of your own, both through ill-measure. If out of your fullness you tempered their want, you would save two lives. Why do you thus gorge your own body with excess, and waste that of the poor with want? Consider what becomes of food—into what it is changed. Are you not disgusted at its being named? Why, then, be eager for such accumulations? The increase of luxury is but the multiplication of filth, for Nature has her limits, and what is beyond these is not nourishment, but injury and the increase of ordure.

"Nourish the body, but do not destroy it. Food is called nourishment to show that its purpose is not to hurt, but to support us. For this reason, perhaps, food passes into excrement, that we may not be lovers of luxury. If it were not so—if it were not useless and injurious to the body, we should hardly abstain from devouring one another. If the belly received as much as it pleased, digested it and conveyed it to the body, we should see battles and wars innumerable. Even as it is, when part of our food passes into ordure, part into blood, part into spurious and useless phlegm, we are, nevertheless, so addicted to luxury that we spend, perhaps, whole estates on a meal. The more richly we live the more noisome are the odors with which we are filled."

Again he writes:

"No streams of blood are among them (the ascetics), no butchering and cutting up of flesh, no dainty cookery, no heaviness of head, nor are there horrible smells of flesh-meats among them or disagreeable fumes from the kitchen. No tumult or disturbance and wearisome clamors, but bread and water—the latter from a pure fountain, the former from honest labor. If at any time, however, they may wish to feast more sumptuously, the sumptuousness consists in fruits, and their pleasure in these is greater than at royal tables. With this repast (of fruit and vegetables) even angels from heaven, as they behold it, are delighted and pleased, for if over one sinner who repents they rejoice, over so many just men imitating them what will they do? *No* master and servant are there. All are servants—all free men. And think not this a mere form of speech, for they are servants one of another and masters one of another. Wherein, therefore, are we different from or superior to ants if we compare ourselves with them? For as they care for the things of the body only, so also do we. And would it were for these alone! But, alas, it is for things far worse. For not for necessary things only do we care, but also for things superfluous. Those animals pursue an innocent life, while we follow after all covetousness. Nay, we do not so much as imitate the ways of ants. We follow the ways of wolves, the habits of tigers; or rather, we are worse even than they. To them Nature has assigned that they should be thus (carnivorously) fed, while God has honored us with rational speech and a sense of equity. And yet we are become worse than the wild beasts."

Sakya Muni the Eastern Master, whose teachings have had so great influence upon millions of men, was opposed to all sorts of cruelty as practiced upon the animal creation. The following lines will prove the loving, compassionate nature of the Buddhist teachings, which we might do well to imitate:

"All beings desire happiness, therefore to all extend your benevolence. Because he has pity upon every living being, therefore is a man to be called holy. Hurt not others with that which pains yourself. Whether any man kill with his own hand or command any other to kill, or whether he only see with pleasure the act of killing, all is equally forbidden by this law. He came to remove the sorrows of all living things. I will ask you if a man in worshiping sacrifices a sheep and so does well, wherefore not his child . . . and so do better? Surely . . . there is no merit in killing a sheep!" (Addressed apparently to the Sacerdotal order.) Our Scripture saith: 'Be kind and benevolent to every being, and spread peace in the world!' The practice of religion involves as a first principle a loving, compassionate heart for all beings. "Hear ye all this maxim and, having heard it, keep it well; whatsoever is displeasing to yourselves, never do to another. In this mode of salvation there are no distinctions of rich and poor, male and female, priests and people. All are equally able to arrive at the blissful state."

Of the rich, Clement says: "They have not yet learned that God has provided for his creatures (man, I mean,) food and drink for sustenance, not for pleasure, since the body derives no advantage from extravagance in viands. On the contrary, those who use the most frugal fare are the strongest, and the healthiest, and the noblest, as domestics are healthier and stronger than their masters, and agricultural laborers than proprietors, and not only more vigorous, but wiser, for they have not buried the mind beneath food. Wholly unnatural and inhuman is it for those who are of the earth fattening themselves like cattle to feed themselves up for death; looking downward on the earth, bending ever over tables, leading a life of gluttony, burying all the good of existence here in a life that by and by will end forever, so that cooks are held in higher esteem than the tillers of the ground. We do not abolish social intercourse, but we look with suspicion on the snares of Custom, and regard them as fatal mischief. Therefore, daintiness must be spurned, and we are to partake of few and necessary things. Nor is it suitable to eat and drink simultaneously, for it is the very extreme of intemperance to confound the times whose uses are discordant."

He also says: "We must guard against those sorts of food which persuade us to eat when we are not hungry, bewitching the appetite, for is there not within a temperate simplicity a wholesome variety of eatables—vegetables, roots, olives, herbs?"

Luigi Cornaro denounces the waste and gluttony of the dinners of the rich. "It is very certain," he begins, "that custom, with time, becomes a second nature, forcing men to use that, whether good or bad, to which they have been habituated; and we see custom or habit get the better of reason in many things, though all are agreed that intemperance is the offspring of gluttony, and sober living of abstemiousness." He thus apostrophises his own country: "Oh, wretched and unhappy Italy! Can you not see that gluttony murders every year more of your inhabitants than you could lose by the most cruel plague or by fire and sword in many battles. These truly shameful feasts now so much in fashion, and so intolerably profuse that no tables are large enough to hold the infinite number of the dishes—those feasts, I say, are so many battles. And how is it possible to live amongst such a multitude of jarring foods and disorders?"

Cornaro's life was one of temperance, and throughout his works he teaches the necessity of a strictly natural diet, pure thoughts and a righteous and noble life.

Tertullian has said, "Nature herself will inform us whether, before gross eating and drinking, we were not of much more powerful intellect, of much more sensitive feeling, than when the entire domicile of man's interior had been stuffed with meats, inundated with wines, and fermenting with filth in the course of digestion, turned into a mere temporary place for the draught."

The children of Israel, who preferred the simple pulse to the splendid dishes from the king's table, became, because of their magnificent physiques, good health and mental power, rulers over the kingdom.

Paul in his teaching points out the beauties of a temperate life, and enjoins all to abstain from meats, and the early Christians were of simple habits.

So we might continue adding to the list of names men whose influence has been unbounded in shaping individual lives and human destinies through the ages. We would write upon our scroll the illustrious name of Sir Thomas More, who pronounced in no uncertain tones against the prevailing drunkenness and gluttony; of Alexander Pope, who, writing of the Future, says:

"No murder clothed him, and no murder fed.
In the same temple—the resounding wood—
All vocal beings hymned their equal God,
The shrine, with gore unstained, with gold undrest,
Unbribed, unbloody, stood the blameless priest.
Heaven's attribute was universal care
And man's prerogative to rule but spare.
Ah, how unlike the man of times to come—
Of half that live the butcher and the tomb!
Who, foe to Nature, hears the general groan
Murders their species and betrays his own.
But just disease to luxury succeeds,
And every death its own avenger breeds:
The fury passions from that blood began
And turned on man a fiercer savage—man."

We would write down that of Pierre Gassendi, of Philip Dormer Stanhope, of that great French pen-painter, Voltaire, of Shelley, most brilliant of all the moral and intellectual luminaries, who have bequeathed to after times "thoughts that breathe and words that burn," for none deserves more reverence from humanitarians than this "poet of poets," who shines foremost among the "Sons of Light." We would number among them Alphonso De Lamartine, Schopenhauer, Thoreau and Wagner. In fact there is hardly a man who has to any great extent influenced public life that has not advocated a fleshless diet, and whose writings did not breathe forth a deep sympathy for the poor, tortured creatures upon whose carcasses we were feasting.

No one who has given the subject even superficial thought can ever believe it just to deprive animals of their right to live, in order that man may simply gratify an abnormal appetite and sate his lust for blood. No one who has ever studied the question from a scientific standpoint, and has considered the physical and moral effects produced upon the human race by the killing and eating of this immense number of animals daily, will hesitate to pronounce it injurious in the extreme. All who have studied the history of those nations who were not addicted to the use of flesh, and have compared their national life, their progress, their arts and literature with that of a people "whose god was their belly," and who spent more thought upon their food than upon national greatness or individual development, have no difficulty in pronouncing in favor of a perfectly natural food as against a flesh diet.

Man has arrogantly inferred that all the world with its abundant animal life, its rich vegetable products and its varied climate was created solely for his own use. This arrogance Pope finely rebukes:

"Nothing is foreign—parts relate to whole;
One all-extending, all-preserving soul
Connects each being, greatest with the least—

Made beast in aid of man and man of beast:
All served, all serving—nothing stands alone.

. . . .

Has God, thou fool, worked solely for thy good,
Thy joy, thy pastime, thy attire, thy food?

. . . .

Is it for thee the Lark ascends and sings?
Joy tunes his voice, joy elevates his wings.
Is it for thee the Linnet pours his throat?
Loves of his own and rapture swell the note.
The bounding steed you pompously bestride
Shares with his lord the pleasure and the pride.

. . . .

Know Nature's children all divide her care,
The fur that warms a monarch warmed a bear.
While man exclaims, 'See all things for my use!'
'See man for mine!' replies the pampered goose.
And just as short of reason he must fall
Who thinks all made for one, not one for all."

Our beautiful, dumb friend, as he comes from the hand of Nature. gentle, affectionate and happy—as he deserves to be.

Our poor, speechless fellow-creature, the victim of man's selfish ness and villiany, his latter years filled with suffering and his death-one of torture and ignominy. Shall these things continue?

CHAPTER X.

OPINIONS OF MODERN THINKERS.

THIS question of diet has had the earnest study of the most learned of every age, and is at present attracting the attention of the masses of mankind all over the world. With the great impulse lately given to the cultivation of man's higher forces has come greater knowledge of the needs of his physical nature, and he has awakened to the fact that in order to attain to the highest state of civilization he must be temperate in eating and drinking. It is not surprising that a question of such vital importance to the human race should be of absorbing interest to all who desire to rescue man from the bondage of his sensual nature and lift him to a nobler plane of life.

Dr. A. O'Leary the popular lecturer and author of that much-talked-of book "Demology," in a private letter to the author says:

"DEAR DOCTOR:—My experience as a vegetarian is perhaps summed up by the declaration that it is so satisfactory that as long as I retain my faculties I shall never allow another morsel of flesh to pass my lips. I have found a better way, and have not the slightest desire to return to the old.

"And yet, situated as I am, constantly on the road, forced to eat at public tables, where flesh-foods are served almost to the exclusion of vegetables, I find it very difficult to follow strictly the diet I should like. Vegetables are usually cooked with milk and butter and other animal fats, and it is almost impossible to obtain a meal without eating more or less grease of some sort. The abstainer from flesh is liable to leave the table hungry while others feast.

"Twice in adolescence and once in early manhood I was at death's door. I was never strong. My mother died at thirty-six and my father at forty-three. I inherited scant vitality and found it necessary to hoard it or perish. I studied works on hygiene, and was forced to the conclusion that food was of the first importance, that health and disease are largely a matter of diet. I therefore determined to eliminate pork from my diet. I noticed no improvement. I quit the use of meat and fish but continued the use of milk and butter, thinking it would not make any difference so long as animal life was not sacrificed in obtaining the food. I observed no great difference in my condition and finally gave up the use of fats of all kinds.

"On June 20, 1886, I ate a mountain trout in Colorado and since then have abstained from flesh entirely, also from all animal fats, except occasionally a little milk on oatmeal, rice or fruit, and sometimes a little vegetable soup, in the preparation of which milk has been used. I notice that after partaking of milk even in such small quantities, I am thirsty and feverish.

"I presume I hardly merit the name of a strict abstainer, as I suppose the dishes I obtain at public hotels are more or less highly seasoned with animal products, though I avoid as much as possible those which contain grease. But what can one do? Potatoes mashed with butter, beans baked with pork, cabbage boiled with beef, the cooking coarse and rank, nothing dainty and delicate as in our homes. Hotel proprietors say their guests demand more and more meat. Vegetables are considered unnecessary; fruit and heavy pastries take the place of natural products which were surely meant for the food of man. I find that since the change of '86 I am stronger, more patient, memory more retentive, passion less furious, sleep more sound, appetite more normal—in short, my physical, mental and moral conditions are more satisfactory in every respect.

"If people would but stop to consider the effects of the diseased condition of the animal-foods with which they are endeavoring to build up their system, they would hesitate long before taking into their stomachs the impurities which too often exist in the slaughtered animals. The fallacy that animal foods are required to make men strong is exploded by the Scotch soldiers, who subsist almost entirely on oatmeal, and the Irish with their diet of potatoes have never yet met their superior—no, nor their equals—on the battlefield.

"Much of human misery is traceable directly or indirectly to the use of flesh for food, which creates an appetite for strong drink, tobacco and other stimulants. The great number of suicides, murders, fights and costly, cruel wars are the result of the same practice. The human race is weak from its continued worship of the 'flesh-pots.' It has been argued by those who favor meat-eating that flesh-eaters are the conquerors of the world. We would point you to the American Indians, who while almost entirely carnivorous yet were conquered by those whose habits of life were necessarily more pure than their own. The strongest of the animal kingdom are those which subsist upon vegetables.

Those which are most useful to man eat nothing but herbs.

"The people who have attained the most advanced state of civilization and whose ideals of life are still the highest and noblest to which mankind aspires were of simple habits, and those who to-day live the most temperately are the most advanced in civilization. One of the greatest impediments of human happiness and human progress is the gross habit of dining off the flesh of our fellow-creatures. Does not the fact that so many religious teachers and religious creeds set aside some days which must not be profaned by meat-eating, and prohibit the use of some animals as articles of diet, prove that within man exists an inherent belief in the uncleanness of animal flesh?

"It is true that he who would walk the mountain tops must walk alone, for few have the courage and the strength to walk by his side, though many will eventually follow in his footsteps."

The illustrious Dr. A. M. Ross, the great English writer whose tongue and pen have done so much to free men from the sensual appetites of their perverted nature, declares that the whole internal and external structure of man clearly indicates his natural unfitness to live upon flesh. Believing this and that the horrors which are heaped upon human beings is the result of a gross diet and filthy habits of life, he has gone on proclaiming "unpopular truths" to the people despite persecution and calumny. He has preached the gospel of cleanliness and temperance faithfully, and his works have been a mighty power for good.

He says it is the first duty of a physician to instruct

the people in the law of health and thus prevent disease, while the tendency has ever been towards a conspiracy of mystery and humbug. Instead of being taught the importance of personal and municipal cleanliness they have been taught to rely on drugs, vaccination and other unscientific expedients.

Says Dr. Holbrook of New York City:

"In a large acquaintance with vegetarians, I have never known one to be a lover of alcoholic drinks or tobacco, and they suffer less from disease than flesh-eaters."

The Jews, who are known everywhere as the most intellectual, the most refined, the most healthful race of the world, will not pollute themselves with the flesh of swine, and by fasting oft, by abstaining from meats as a common article of diet, have always been, even as they are to-day, a nation clean in every respect. They do not boast of their strength, but "still waters run deep," and in every country the Jews, despite the opposition, the persecutions, the bitter hatred of the people among whom they have been forced to live, have kept sacred their laws and traditions, and are the strongest, healthiest, wealthiest and most powerful people on the globe. They control the capital of every nation, and while their name is "a by-word and a reproach," it is also a power and an honor. "I am a Jew" is spoken as proudly to-day by the sons of Abraham as in the days of Israel's glory.

Mrs. Ellen Eames DeGraff, who has done much to propagate the high ideals of which so many but dream, speaks thus of the birth of her youngest son:

"Before the birth of this child I had conceived the idea of a child born under harmonious conditions, and had thought much of what its future might be. I was during this period actively engaged in philanthropic work and in philosophical studies. I gave the subject of cheap living such careful consideration that I was afraid my child might be miserly in disposition; but instead he is careful, not stingy, and I feel that the other studies which engaged my attention at that time have helped develop other traits and thus make his nature harmonious.

"I read and studied the writings of Count Rumford, and hope that the thoughts I have gained from his works will yet mould the life of my child into a noble, well-rounded manhood.

"During gestation I subsisted upon cereals and fruits, and after his birth ate largely of oatmeal and cream. For my first child I had plenty of milk until she was a year old, but for those which came after I never had sufficient milk until the birth of this boy, for whom I had enough for a year. When my baby was born he was a lovely child, and has continued to this time a most beautiful boy, of quiet, loving disposition, studious in his habits, of a philosophical turn of mind, with an intellectual and spiritual nature and with a strong taste for studies of a serious nature."

Regarding her personal experience as a vegetarian, Mrs De Graff says further:

"I can not speak for others, but my own development in higher thought is utterly inconsistent with meat-eating. I shudder when I see my innocent, tender-hearted, brown-eyed little vegetarian boy looking on at the *post-mortem* of a respectable, harmless hen, whose body, filled with eggs, is such a pathetic comment on a useful life, full of promise of motherly faithfulness, cut short by the murderous hand of one of the neck-wringing species—and try as I may, I can see nothing but speciousness in the clever arguments which meat-eaters adduce to prove the ethical correctness of their position. Benjamin Franklin was a vegetarian from principle, until one day seeing a fish for which he longed cut open, he beheld other fishes inside. 'Oh,' said he, 'since the fish eats his fellow-creatures, I will eat him." Of course if we do not aspire to a higher conception of our duty to our living companions than does the brute, we are absolved. But these very reasons defeat their own arguments by stating that we are kings of creation, and all things are for us! If so, a burden of responsibility in regard to our duty lies upon us, and we must be more considerate and human than even a sheep, the mildest of all created creatures.

"To be brief and sum up my present belief, I will say that I believe non-meat-eating to be best. In its place I believe nuts and fruits should be used, for science has placed us by virtue of our formation among the frugivora. I do not think any one should be in any degree coerced into the adoption of a non-flesh diet. As long as it seems right to my brother to eat meat, it is right. My aim is only to encourage those who long to be free, by assuring them that they will be at least as well off without it as they were with it. Were it not so Nature would have entered her protest by a longing which could not be overcome. If I longed for meat I should eat it.

"Spiritual growth is quite apt to remove the desire for animal food, although it would be very sweeping to claim that it always does. Some of our grandest philosophers and best and kindest men have been meat-eaters. I have known people to gradate from vegetarianism to meat-eating. I do not believe cooked vegetables to be natural to man, and it does not surprise me that one rests upon one tight-fitting shoe to ease the other. I hope some day to see the fruitarian child, as I have seen the child of the cereal food."

Mr. Edward Martin, a well known artist and photographer of this city—the designer of that remarkable Biblical picture "The Arch of Life"—who is a warm personal friend of the author, in a recent conversation said: "I had long been cognizant of the fact that you were doing an immense amount of good, and had often thought I would like to adopt a different diet, but I did not know how to begin, and as I had been taught to believe meat was necessary for man's existence, I did not think it possible for him to do without it entirely and be in perfect health. That trip we took together was the beginning of a new life for me, and I could not be tempted to return to the old habits."

The trip to which he referred was taken several years ago, when we visited all the principal cities of the East. On this tour Mr. Martin determined to give my system of diet a trial, and to eat the same things I did. When I had given my order to the waiter at the hotel, he would simply say, "I will take the same," and thus by imitation had proven to him the fallacy of the objection so often offered—"I can not obtain proper foods at the hotels, boarding houses and restaurants, and am not able to live as I should like." It is often difficult for the beginner to know just what kind of food to eat, but a little thought will enable him to select a variety of vegetables and fruits, corn, Graham or whole wheat bread, and nuts, cereals, etc., to satisfy perfectly every demand of Nature.

When we started East Mr. Martin was in bad health, and had been for a long time. He was, of course, skeptical as to the benefits he would derive from the diet, but was determined to give it a trial while with me, in order that he might have my advice, and as he had long desired to prove for himself the efficacy of the principles I advocated, knew he would never have a better opportunity of doing so. When we returned home a few months later he was very much improved in health and spirits. He had no desire for flesh, and when meat was placed before him, found it impossible to partake of it. He pushed it aside, remarking to his family, "The rest of you may eat what you please, but no more meat for me." He has since that time never tasted meat, and finds that even the sight of it nauseates him. He has also lost all desire for tobacco, which now has the same effect upon him as meat.

To-day Mr. Martin is in perfect health, rosy-cheeked and robust, and is a striking contrast in every way to the man of a few years ago. He is happy and buoyant, for the gloom of dyspepsia no longer broods over his life; he can accomplish more work and better work than ever before, for he finds an inspiration in life which he had hitherto lacked. He is no longer haunted by the melancholy fancies of a disordered brain produced by a disordered stomach, but his whole physical system is pure, and consequently the functions of his mental organism are perfect.

The benefits that have been derived by Mr. Martin from a change in diet, and the observation of the simple laws of Nature, may be realized by all who will give them a trial. There is no possible need of people being miserable and sad, melancholy and depressed, when they may as well be full of the richness of life and know the joys that come from true living. With all honesty I must say that if people are sick it is their own fault and they deserve no sympathy.

Alice B. Stockham, M. D., a popular writer and an earnest worker in advanced lines of thought, whose name is doubtless well known to many of our readers, in response to a letter addressed to her upon this subject, writes:

"DR. L. H. ANDERSON.

"*Dear Sir:*—In reply to your request for a word from my pen for your book on Natural Diet, you have my cordial permission to use any matter from the enclosed articles. While I do not consider meat an *unhealthy* diet as many do, yet, regarding the subject from the æsthetic point of view, I can not believe it is either physically clean or morally right.

"It seems to me that when we realize the unity of all life, when we understand that life, love and intelligence pervade all Nature, that we are one with the Divine creative force, we will be very slow to take conscious life for our sustenance. Of course I know it is difficult to know at what point in the products of Nature that sentient life begins, still it seems very plain to me that the domestic animals that we love and that evince in such a remarkable way their affection in return for our care, should certainly be included under this head.

"I think it has been well proven that animal food is not necessary for sustenance. We have all the grains, vegetables and fruits, including a great variety of nuts;

and out of these every element required for the body can be procured, and that, too, in great abundance.

"I would make a strong point of using nuts as food. We have been inclined to take them as we would condiments or dessert, but I should recommend that they constitute one of the chief staples of diet. People who complain of the indigestibility of nuts are those who use them only after they have partaken of other foods or between meals. I prepare nuts in soups, in croquettes, in salads and in hash. All of these preparations can be made very delicious, more so than if they were prepared from flesh, and are really more nourishing. Of course I also place shelled nuts upon the table frequently.

"We should aim to simplify the preparations for the table. One of the growing failings of the nation is the complexity of diet. In the study of sociology it comes to me very strongly that one great point that should be made is furnishing a diet that requires less labor, to say nothing of expense. While neither of these things are obtained at first by substituting the natural diet for the flesh diet, yet one learns in time, when the nutritive values are understood, to simplify the preparation of food. Combinations of the different materials add nothing to their nutritive value and little, if anything, to their palatableness. I trust that you will be able to point this out in your work, and believe that you will do valuable service in many ways in your forthcoming book. With kind wishes, I am,

"Sincerely yours, A. B. STOCKHAM."

Dr. Stockham during her travels in the East found that while in some instances the Buddhists use flesh as an article of diet, the practice is not prevalent. In her address "Food of the Orient," delivered before the World's Fair Congress, Chicago, June 8, '93, she says:

"It is true, I think, as we have been led to believe, that Buddhists as a whole have a great respect for life; the provocation must be very great indeed if they take it. Insects, though obnoxious, are never killed, while birds and animals come and go with the greatest freedom. It is possible that the introduction of animal food in their diet is due to the advent of other nations.

"It is nearly eight hundred years since the Mohammedan invaded India. He eats all kinds of meat except pork. It is not so long since the Portuguese settled on the western coast of that country. It is over three hundred years since the Dutch in great numbers became dwellers in the seaport towns of Burmah and Ceylon, and have insidiously influenced all the customs and habits. Then, later, the English, with their beef and beer, have taken possession of all these countries. The wonder is that these susceptible people have not more generally adopted the barbarous habits of the Occident.

"The Brahmin and all the Hindu castes below him, if they are true to their customs and traditions, never use animal food. They too have a great respect for conscious life, and feel that their own lives must not be preserved by taking life. However, so far as I could observe, their greatest and most potent reason for adhering to a vegetable diet is their great abhorrence to dead animal matter. As soon as life leaves any created thing, that thing to them is polluting, and no power except fire is great enough to eradicate that pollution. The corpse of the dearest friend or relative becomes at once obnoxious, and any form of a carcass or any part of the same

ALBERT H. SNYDER.

(See page 224.)

is pollution itself. The hair, the skin, the hoofs, the feathers, the fat, all are simply obnoxious to the Hindu. You can not imagine the disgust with which he looks upon a western lady who wears gloves made from the skin of a dead animal. Leather can never enter into the composition of his shoes, nor hair nor feathers into the bed upon which he sleeps. He will not even use soap made from the fat of a dead animal. This feeling of repugnance is born in him through ages of teaching and traditions, and can we wonder that he calls the meat-eating and soap-washing people of the western world 'filthy barbarians?'"

Speaking of the people of India, she says:

"All Hindus are not stalwart men, but the Brahmins who, as I have shown, are most abstemious in their living, and must from the fixed laws of their caste adhere to a rice diet, furnish samples of as fine physique as any people in the world. They stand straight, are broad chested, lithe and supple of limb, and in no way give any impression of weakness.

"The men from Punjab and Rajputan, northern provinces of India, are noted the world over for fine physiques, for strength and endurance. These are provinces from which the English draw largely the native soldiery. These are the men who are seen standing as invincible statue-guards in the English treaty ports of China and the islands of the East.

"Going from the Pacific coast to the Orient, this soldier gives you the first picture of the Hindu. In his red turban, his simple uniform of American jeans, stalwart and implacable, he at once gives the lie to all the romances of India's dependence and weakness.

"The Brahmin as a class proves that man does not

live by bread alone, nor rice either. His whole life is one of training to make the body serve him—the master. Not the body only, but every desire, every ambition, every self-interest is denied away until he stands alone with Atma-Buddhi, until he is at one with the creative principle. This is not only true of the Brahmin, who spends hours daily in contemplation and repetition of mantras, the substance and power of which is to effect freedom from the things of sense, but he being twice born and favored of the gods is held as an example by all castes beneath him, and thus all Hindus come to feel that to be a slave to bodily wants, to physical necessities, is not only deplorable but despicable.

"You can not imagine what barbarians we Westerners are considered, we who are devoted to flesh-pots and beer-mugs, and who can not travel without bath-tubs and sponges, and in preparation for days of hunting and sight-seeing do not even forget finger-bowls, tea-baskets or air-beds.

"There is a mighty lesson that the English could and should learn of the Hindu, but do they, and would we Americans, were we in India? We Western people are so sure that progress means science applied to the physical world, and we are so trained in the thought that the objective world is the only world, is what we live in and for, that we are mastered by it and its condition. This hand, this arm, this head is *the man;* this outward man must be fed, clothed and protected, and through this belief we come to know no other world—no other man.

"Be not dependent on any one material substance. Have no musts in your bill of fare. Live, live the higher life; compel the body to obey you, for you must know that you may be the master. If you, like the Anglo-

Indian, have not your pepper ground on the table, or miss your *chotohowzer*, you still may be happy. It is yours to enjoy the glories of a free life on the prairie or on the mountain top, and it also may be yours to rise to great spiritual heights, Bunyan-like, in a dungeon.

"If through any process you come to recognize the power and dominion of the spirit—are 'born again' or 'twice born,' as the Brahmin—you no doubt will choose fruit and grains for your food; you will not wish conscious life to be taken for your sustenance.

"The inconsistency of the man who one day exhibits great fondness, indeed almost adoration, for a pet animal, and the next day becomes a butcher of this life, or a similar one, to subserve his unnatural appetite, must cease.

"You will come to reverence all manifestations of life and know that the bird and beast have the same right to it that you have.

"Many through a knowledge of a higher life, through spiritual perception, have learned to abjure meat; it becomes repulsive to them and seems inimical to the higher development. This is notably the experience of many Theosophists. Count Tolstoi renounced meat when he learned of the Christ-life, and set his feet in the way of righteousness—right living and right thinking.

"Let us remember that there is a spiritual law, and if we understood it, if we are trained to become acquainted with its workings, to let it have dominion over us, the needs of our bodies will not dominate us.

"Henceforth none shall spill the blood of life, nor taste of flesh, seeing that knowledge grows, and life is one, and mercy cometh to the merciful."

If this advice were followed a great deal of unnecessary work and worry would be saved the people of our Western world. When we fully realize that the body, as a machine, takes but little material to keep it in perfect repair and in good working condition, we will have learned a most valuable lesson. Instead of gorging the stomach with unnecessary foods, which obstruct the perfect functions and make them unfit to perform their proper duties, we will by temperance keep them clean and pure and fit for the highest work of which they are capable.

In her article on "Artistic Living," Mrs. Stockham says:

"Artistic living includes the choice and preparation of foods," and after dwelling at some length upon the interior forces of man's nature and his advancement along the lines of higher spiritual life, expresses the belief that when man shall have arrived at a true conception of the real beauty of life he will concern himself less about his bodily needs and more regarding the needs of the soul. "Undoubtedly," she says, "a natural selection in diet will be largely from the grains, fruits and nuts. He will also simplify the preparation of these, eliminating the evidence of drudgery that usually accompanies the complex *menu*.

"In table-living we have become as much a slave to the conventional as we have in dress. We give the ten, seven or five-course dinners, not because it expresses our individuality, not to supply the needs of the body, not because it is according to our best taste, but simply because it is the fashion.

"Our wealthy neighbors have been to Europe and find that service is cheap, the *table d'hote* is in common vogue with people of means, so he adopts it in his family probably as a means of displaying his wealth.

"The foolish fashion-follower, whether he has wealth or not, with one servant or none, falls into line and also serves from three to five courses. He demands this, forgetting that the most energetic and capable woman, with one or even two servants can not consistently inaugurate this custom without a defection from justice. By strain of purse and pulse he sacrifices living art to imitate his neighbor, to be and do as other people, instead of being his true self.

"Table decorations appeal to taste as well as to the eye. A salad that looks crisp and cool gives an anticipation in flavor. Coffee that has a golden tint and the aroma of Arabia has no discounts when presented to the palate.

"In china-ware harmony in color and adaptation to purpose lend attractiveness to food, while artistic arrangement of flowers and fruit contributes especially to development of art and soul."

Again she says:

"Man's recognition of the unity of all life, of the one law expressed in diversity compels him to choose that conscious life shall not be taken to supply his table. Birds, beasts and fishes become as fellows—fellow-intelligent life-manifestations, endowed with rights demanding respectful consideration. Strange to say, the unfolding artistic sense of taste soon rejects with an abhorrent shudder all the cooked products of the flesh of their fellows.

"The cook's art, with all his inventions of chemical compositions and even curry seasonings, fails to conceal the meat flavor so as not to be discovered by one who has forever discarded flesh as food. It is to him like the abhorrence of the Hindu for a glove made from the decayed skin of a dead animal, to be worn upon the living, rosy flesh of a lady's hand. It is to the artistic sense the neglected, reeking, decaying outbuildings of a nobleman's villa.

"It is to him the polluting corpse, the partaking of which may not contaminate his body, but shrinks and dwarfs the soul's perception of things.

"The partaking of flesh is a barricade to the evolvement of the Divine principle, to the shining forth of the soul's intelligence.

"As every injustice of man to fellow-man reacts in degradation and dulled perceptions of the perpetrator so does the countenancing of taking life for food react upon a nation's progress.

"Artistic living brings larger experiences to the soul, it traverses heights and depths hitherto unknown. *All matters become but a manifestation of mind*, all physical conditions thought expression."

A beautiful thought and one that will find an echo in the heart of every seeker of the higher, purer life. Let man study carefully this great subject in all its bearings and he will never again wish to deprive of existence the dumb creatures whose life is as sacred as his own, nor pollute his lips with the decaying carcasses of dead animals.

Dr. Stockham in her book "Tokology" which has been a blessing to so many mothers, advocates a vegetable diet for the pregnant woman and the subject of pre-natal culture considered in relation with this becomes full of possibilities. Many women testify to the value of a fruit and vegetable diet during those critical periods.

The subjects of a pure diet and the sacredness of animal life are being discussed widely by the thinking public. The following by Mr. C. M. Loomis, clipped from a Chicago paper of May 22, 1898, shows the trend of thought upon these important issues. We are sorry not to be able to quote the whole article, but have culled the most salient points:

"Though at the present time the meat-eaters preponderate in multitudes, there is a steadily growing sentiment against flesh-killed food little dreamed of by those who are uninformed. Dr. Kellogg is said to have been cured of meat eating by seeing the cook cut into a full-grown abscess secreted in a quarter of beef. Other noted vegetarians stoutly declare that man should not feed upon the same kingdom to which he belongs; theoretically that he is man, a creature endowed with attributes above the animal. If he is manifestly above the beasts and birds, and can subsist directly upon the generous abundance of the earth—as thousands have proved by actual experience—must there not be some forthcoming rescue for us from the time-worn habits of our fathers?

"It is a faulty parallel, and a meager credit to man, to suppose that he must subsist on flesh-food because the lower animals do the same. My experience has proven that it is not at all necessary. Here is my one important reason for abstinence from meat: The instant the animal is killed the refuse matter throughout its organism, and which is on its way to be excreted, is stopped, and, of course, the impurities in each fleshy cell goes into the human stomach for better or worse—some intestines to receive and cast off said impurities, others to permit the taint to find its way into the blood.

Incidentally I am of the opinion that all flesh eating is a mistake, handed down to us along with many other errors of the past. But be this as it may, a dozen years of abstinence in my family has served in an incalculable degree to encourage cleanliness of body in other ways, until now it would be a sacrifice of both our moral and physical welfare to eat of that which we no longer crave. Ask the vegetarian if it is not more God-like to take his food in all its purity direct from mother earth—ripened in the sun and free from animal taint, and he will smile that such a question should even suggest itself.

"The stomach chemicalization of the beast need not enter into human necessity. Already we have a cooking fat made from cocoanut and cotton seed far superior to lard, wholly vegetable and quite as cheap. It is said that science has produced a sole leather made wholly from wood fiber, and were we to be deprived of the hides of beasts, there would soon be found a way to produce ample foot coverings by scientific discovery. In other words, all that is contained in the animal organism can be gathered from vegetation, combined, chemicalized and transmuted to supply our needs, and because of our discoveries, doubtless come to us in far superior qualities. I believe we have only an inkling of the wonders of science. If certain properties of nutriment are contained in the cow's milk, were the supply stopped necessity would soon demand, and there would be produced, a carefully prepared substitute for the babes. True, there is a warming, wholesome companionship between the farmer boy or girl and the dumb creatures upon the farm, and I say that to foster that I would use my every effort—foster it by omitting the killing thought entirely, and thus observe the commandment "Thou shalt not kill" in all its simplicity of diction.

"We are in a great cycle of refinement. Many examples are there of people who have progressed beyond the habit of flesh diet and do not know it, and who are suffering from dyspeptic ailments in consequence. Let them put aside the old habit for one year, and I dare say that in a large number of cases the unwelcome ailment will take flight.

"It was only a few weeks since that our city papers chronicled the case of a pig-killer at the Stock Yards going stark mad, so that it took several policemen to overpower him. Is not all this consistent with the man's calling? Callous as may be the human perceptions, somewhere in the butcher's consciousness is a pang of remorse for every act of killing. We have no right to take a life which it is not in our power to restore. We go out and slaughter the beast—not in anger or in self-defense, but deliberately—and to that flesh food has been transmitted a murderous taint. Our son eats of that flesh and perchance goes out and kills a neighbor. Should we wonder at this? With brain finely organized, he may be thrice susceptible to the taint we have put into the flesh, and unable to master himself he commits the crime in a moment of irresponsible frenzy. Ask him why he did it and he will tell you he does not know. His parents are the ones to blame for the atrocious act.

"Cease raising the bovines and the hog, and when the superfluous demand does not exist Nature will kindly regulate the supply.

"But we have need to be most tolerant, since we are aware that only those who have become emancipated from the meat-eating habit know the freedom thereof. I mean truly emancipated, not acting under the lash of duty without conviction. There is less passion coursing the veins of vegetarians; they are cooler in summer and warmer in winter (I speak from personal experience); they think better, sleep better, have better health, while appetites for strong drink, so prevalent with excessive meat-eaters, is unknown among them.

"I would not dare, with my present views of right and wrong, to give pain to a single creature. Since I put aside meat diet my attitude toward the animal has changed. I see in the great mastiff's eye the devotion of a human saint, and in the face of the motherly cow a trace of sublimest affection. But when I was pursuing the dumb brute to kill him I saw not these things. I was blinded by the race thought which commands us to do as our ancestors did before us, regardless of conscience or sentiment.

"'But,' says the meat-eater, 'you are taking life unconsciously every breath you draw.' True, so we are. But unconscious are we then of a motive to kill; hence there is no violation of the moral code. It is when a man says, 'I must slay this creature; its wonderful organisms must be torn asunder and the God-given life-essence be driven out that my hunger shall be appeased'—an act of violence, mind you, back of which must be the thought of the destroyer—that the wrong becomes manifest. What is the reverse of this? The abstainer says, 'No pain shall I inflict, no thought of taking life shall possess me, and on the most nutritious foods given me by Nature will I subsist.' Truly there comes into the soul of such a one a peace which transcendeth all things.

"In my immediate family the transformation has been

complete. My daughter, now nearly 13, has grown up without meat, and has a most remarkable vitality and mental aptitude. We have no desire for nor do we use tea or coffee, tobacco nor stimulants; we keep no medicine chest, nor have we a family physician. Before the change from a meat diet I was the victim of ill-health, and was scarcely out of pain a moment. For years past we have had health in abundance. The time has gone by for the belief that vegetarians must necessarily be a cadaverous looking people. I only refer to the above to show that what I have written is not merely theory, but the results of actual every-day experience.

"I rejoice that with no boasted skill whatever, and by following the mere instinct with which my stomach and moral sensibilities are endowed, I can get along with one-half the food consumed by the meat-eater, have no taste for condiments, very little for sweets, and none for tobacco or liquor, and yet keep my weight, my health and my temper, and, in a measure, woo to my aid that elixir of youth which the learned alchemist of the past has sought to invoke by mechanical transmutation. It is surprising when we come to know just what a normal appetite is. When food is properly assimilated, few delicacies placed upon the table, and when a natural stomach craving comes to us regularly at meal-time, only a merest amount of food is necessary. It is our attempt to depart from simplicity that damages our victuals. Our back-of-the-ears propensities and animal passions are increased by a flesh diet. Necessarily, then, the habit is a gross one. Witness the inflamable temper of the tiger, the lion, the hyena and other carnivorous animals. Then recall to mind the docility and wonderful endurance of the camel, the remarkable memory and unwearying kindness of the trained elephant, the faithfulness and fortitude of the horse or ox, the sleekness and beauty of the deer, the harmlessness of the sheep or goat—all herbivorous animals. Can we ask for better evidence as to the degrading tendency of a meat diet? If flesh diet acts thus upon the beast, so it must in some degree upon the individual who humors his appetite for flesh. All the efforts and preaching of temperance advocates will do no good so long as meat is eaten. Meat is a stimulating food; with some temperaments it creates a taste for stimulating drink. A sad verdict this, since so much time, money and prayers have been spent in the drunkard's behalf. Has any person ever discovered a saloon within the boundaries of a vegetarian community? The very thought is incongruous and illy matched with the sobriety of meat abstainers."

Prof. Uriel Buchanan writes the author:

"The organic quality of man corresponds to and is a resultant of the kind of food used. The nervous activity and the animal passions are increased by a diet of flesh, while a diet of vegetables, fruit and grain produces scientists, scholars and thinkers, who are mild and gentle, aspiring and exalted.

"The man who takes his food in all its purity direct from Nature's bounteous store, instead of eating the tainted flesh of murdered beasts, is in closer harmony with Nature and all living things than he whose existence is maintained by the sacrifice of the helpless dumb creatures of earth. His brain cells respond to the impulse of nobler thoughts, his character is more beautiful, his soul more pure.

"This is an age of refinement. The savagery of

former times is passing from the heart and soul of man, and he begins to see deep in the eyes of the animals the struggling fragments of divinity. He sees in the eye of the dog the affectionate, mute glance of devotion, and in the face of the motherly cow a trace of the attributes of humanity.

"For a more perfect physical and mental growth, and a higher unfoldment of the soul, man must discard the diet of flesh, and take his food supply first handed from the storehouse of Nature; for all life is sacred, and to take the life of the least of God's creatures is contrary to the spiritual law of man's being, and is a violation of the divine commandment which says, 'Thou shalt not kill.' URIEL BUCHANAN."

Prof. Ormsby writes, under date of June 15, 1898, as follows:

PROF. L. H. ANDERSON,—

Dear Sir:—In replying to your request for my views on the subject of vegetarianism, will say, I can not be considered a vegetarian in the strictest sense, from the fact that I use animal products to some extent; but the food of that nature that I do use I consider the fruit of the animal kingdom, and, as such, suitable for man, although not really necessary. I have two simple rules which ever guide me in matters of diet, and I think they cover the ground from any standpoint we wish to view the subject. They are as follows:

First: Take nothing for food that Nature does not yield to the hand of man without resistance.

Second: Take nothing into the system as food that does not contain the vital life principle, capable of reproducing its kind when given opportunity for such expression. This latter statement may need some qualifying to make it clear to your readers, but we will only present the idea here, as space will not admit of more.

The idea is this: We must, in order to add life, animation and increased power—especially mind power—and spiritual clarification to our treasures of worth, take such things from the great currents of life as will add to the qualities which are building far more potent and lasting conditions, otherwise we gradually undermine the forces of being and disintegrate until we become a useless animal. The fact that one can not live on meat alone should convince us that meat is not a fit substance for food. But with the whole fabric of thought colored by *dies* that have no permanent base, its hue is in accordance with the heterogeneous mixtures which for ages have been hashed up to gratify the extremely animal nature of the great mass of humanity. A few only really know of the advantages of vegetarianism. It is so hard to make the change. It is like changing an inebriate to a condition of total abstinence for some to leave off meat and readjust their appetites to a purely vegetarian regime.

"The vegetarian cause is worthy of deep consideration, and I believe it is one which will play an important part in the future welfare of the race.

"Sincerely yours, F. E. Ormsby."

Miss Grace B. Moore, publisher of "Social Culture and Laws of Life," and "Moore's Marvelous Memory Method," says:

"Like many others I inherited skepticism in regard to the necessity of a flesh diet; but, also, like many another I was met at every turn with so many apparently sensible arguments in its favor that, being young and easily overruled by those whom I considered competent to advise on the subject, I pursued my usual diet, which included meats, although in my heart I felt it was not right, and that the sacrifice of innocent life was not essential to my development, either mentally or physically.

"It was my good fortune five years ago to become associated, in a business way, with Prof. Anderson, who in many an argument which we held for and against the prohibition of flesh food, and the substitution for same of the ideal diet of fruit, nuts and vegetables, successfully combatted and overcome anything I was able to present, either of my own or from hearsay, relative to the superiority of animal food. Added force was given his arguments by his own life, which was based on a strictly vegetarian and fruit diet, and recognizing his perfect physical and mental condition I was forced to acknowledge that he knew whereof he was speaking, and was quite competent to advise. I also took note of the many cases which he assumed the care of, and which, with consistency, he led along the lines of this natural way to health and vigor, from a physical standpoint, while at the same time the mental condition was clearer and stronger and thus each one treated was benefited symetrically. Gradually the mists and weak doubts were cleared away, and it hardly needed a visit to the stock yards of Chicago and the watching of the murdering of helpless animals, whose lives were sacrificed to cater to the cannibalistic tendencies of their human brothers and sisters, to completely convert me to the more humane mode of living as evinced by the reform diet so-called. As I watched those poor dumb animals led to the slaughter and beheld their trembling and appealing looks to be saved from such a fate, realizing that they knew exactly what was coming to them, but having no way of communicating their agony of fear but through their great limpid eyes, and, as I saw those eyes turned as if begging me to intercede for their lives, I realized all the piteousness of the whole transaction, and it did not need the beholding of the unnecessary torture of those naturally gentle creatures in the way of prodding with sticks, beating, etc., to waken in me the thought "Mea culpa," for was I not one of the many who cried for the sacrifice of these helpless creatures, and was I not then guilty of their blood? Loose fell the shackles with which heredity had bound me, and I stood forth in all my freedom, determined that thenceforth no life should be offered up at my shrine, but that, on the contrary, at all times and in all places my voice and influence should be found advocating the ideal diet, as thereby I was releasing these creatures, whose lives are dear to themselves, from the inhumanity of man."

Albert H. Snyder, the well known journalist and President of the Chicago Vegetarian Society, says:

"There are too many influences working against progress. The influences of bigotry, of superstition, of ignorance, are strongly against advancement.

"It will take years and years to get the people out of the rut in which they are living. It may take centuries to lead them out of the footsteps of their ancestors, who had not the same opportunities for the acquirement of knowledge as the present generations.

"We know that most people live as they do, and think as they do, because their parents and their grandparents and their great grandparents lived in the way they are living, and thought the thoughts they are thinking. It has never occurred to them to go higher.

"How much more rapidly civilization would progress if the masses could learn to think and to investigate! The world is in need of investigation along many lines.

"There is urgent need for an investigation of the diet question. The food problem may seem an insignificant one to those who have given it no thought, but viewed from the standpoint of vegetarianism it is a momentous one.

"I may be prejudiced when I say it, but I honestly believe that the universal acceptation and adoption of the vegetarian idea would mean the banishment—indirectly, if not directly—of most of the ailments and troubles which perplex human kind. Drunkenness and cruelty would be unknown. War would be unheard of. There would be less sickness and less poverty. There would be little excuse for the existence of jails, and insane asylums would cease to exist. There would be a general observance of the golden rule."

The Rev. Henry S. Clubb, minister of the Bible Christian Church, Park Avenue, Philadelphia, has a fund of material from which to draw an extremely interesting and somewhat romantic biography.

Born at Colchester, Essex, England, June 21, 1827, he has just celebrated his seventy-first birthday in the full enjoyment of vigorous health. When a boy he became a vegetarian from hearing conversations between his father and William G. Ward, a commercial traveler, who was an enthusiastic food reformer. He was a skilled conversationalist, and vividly portrayed the horrors and cruelties of the slaughter-house and the danger of eating diseased meat.

Being very sensitive and thoughtful he realized the horrors of killing, and he noticed every time a fatted hog had been butchered the children were invariably made sick, so he drew his own deductions and was strengthened in his determination to eschew flesh as an article of diet. He watched the effect of the change in his food, and observed that after partaking of a meal in which flesh formed a part he had a coated tongue, a disagreeable taste and dryness in the throat, whereas after a meal of bread and milk, baked apples, ripe berries, etc., no such disagreeable results followed. His later experiences confirmed his early observations of the effect of different foods, and he grew up to be as healthy a specimen of English youth as could be found.

Mr. Clubb emigrated to the United States in 1853, and attended the fourth annual meeting and festival of the American Vegetarian Society during August of that year at the Bible Christian Church, Third Street, Philadelphia, and in conjunction with a committee appointed for the purpose, compiled an address on cholera founded on the fact that in no instance had a vegetarian died of this disease.

During the interim between this time and 1871, Mr. Clubb held many positions of honor and trust, including active service during the civil war. In all this time he was true to his principles in benefiting humanity, always to be found using his pen and influence in favor of the weak and striving by his efforts and example to raise the standard both of mental and physical excellence.

In 1871 he was elected State Senator of Michigan, resigning the office of Alderman to accept same, in which position he promoted the fruit interests of the

State, which is now known as the Banner Fruit State of the Union.

In 1876 he visited the Centennial Exposition at Philadelphia, representing several prominent newspapers of Michigan. This brought him in contact with the Bible Christian Church again, and he was prevailed upon to continue as pastor, which position he has held ever since. In 1876 an attempt was made to renew the Vegetarian Society, which, during the war, had ceased to exist, and these efforts led to the organization of the V. S. A., and in 1888 *The Vegetarian* made its appearance under his management. Later *The Food, Home and Garden* was established instead, and has been maintained ever since with unabated vigor and constant improvement.

As an evidence of the result of sixty years of vegetarian experience, Mr. Clubb said, when addressing a meeting of the Vegetarian Society in New York City, May 20, 1898, that when he went to Manchester in 1848, after living without flesh meats for nine years, he was regarded as a very healthy young man. Mr. C. A. Montgomery arose and remarked, "And you are yet," which was endorsed by the applause of the audience.

Becoming a convert in his tender years and strictly adhering to vegetarianism through all his life, he stands at seventy-one years of age a bright and shining light as to the success of such mode of living. He has been tested in every way, physically and mentally, and has never been found wanting. His life has been largely a public one, and to every call upon his mental or physical powers he has responded readily and intelligently. Always before the public, which is ever quick to seize upon any lapse from consistency as evinced by any reformer, he has nobly withstood the test, and now at this ripe old age can proudly cite himself as an example of the results to be obtained from a natural diet, and there are none who can gainsay him.

It was by mere chance in 1891 that copies of *Food, Home and Garden*, referred to above, fell into the hands of the author of this book, and which made him a convert to vegetarianism.

I can never be sufficiently grateful that I entered upon such a life, as I am firmly persuaded that it is the only natural, and, consequently, ideal diet, and each year but adds to my delight in same, as through its means I enjoy life in its perfection, with none of the hindrances to be met with that so often interfere in the way of indigestion and kindred complaints to mar the happiness of the flesh-eater's existence. And I have the further satisfaction of knowing that no innocent creature has been sacrificed to contribute to my well being.

The future of the race lies in the hands of the mothers, and I sincerely hope that every mother who reads this book will be lead to study foods, their properties and influences upon her life and the lives of her little ones. I trust, too, that it may be the means of leading all from their gross and pernicious habits of eating and drinking to habits of temperance and purity; that those who have suffered from ills, the causes of which they were unable to understand, may come to a realization of the fact that they are themselves to blame, and be led to follow the laws which bring health, and with it perfect happiness.

I have studied the subject so thoroughly, and have watched the outworkings of the law in so many instances, that I do not hesitate to make the statements I have, though to many they may seem absurd. My wide experience as a physician and teacher has given me abundant opportunities to investigate personally among all classes and conditions. I have read extensively upon the subject, have examined the opinions of the most noted authorities of every age, and my conclusions have been the result of many years of patient toil.

So I am fully prepared to defend my position, which has been the same as that held by sage and philosopher since the first dawn of history, and to overcome all objections which may be offered this theory upon any ground whatever.

However, my object in this work, as in all my books, has been to make people think for themselves, to cast aside the chains of custom which hold them in a narrow groove, and compel them to do continually those things they dislike to do, and to dare to be free even though others are willing slaves. I have endeavored to place the subject in such a light that all will feel it incumbent upon them to study it more deeply in its bearings upon their own life and the race at large.

That the cruelties which can not be dissociated from the killing of animals, whether for food or sport, degrade firstly all that take part in them, and next all who profit by them; what these cruelties are amount to a "tale of shame," and that they can co-exist in a country that boasts of its charitable and religious institutions is an anomaly painful and inexplicable.

Dreadful are the revelations made by humane men, who, setting aside personal comfort and peace of mind, have endeavored to sound the depths of animal agony and bloodshed. The process of flaying alive, and even of dismembering animals before the breath has left their bodies, is far from uncommon in slaughter-houses, though the public is not aware of what passes. In the slaughter-houses which I visited recently in company with Dr. W. J. Wiesen, one of Chicago's meat inspectors, lambs were being killed, the process consisting in hauling the wretched animals up by a chain attached to their hind legs and then cutting their throats while they hung struggling head downwards. I saw hogs stabbed in the same manner, and dropped into boiling water while yet alive.

I have endeavored to unveil some of the horrors of the slaughter-pens, and to show that disease and death lurks in the corrupt flesh so often used for foods, in order that every one may be fully aware of the danger of incorporating the bodies of animals into their own physical systems. I have also tried to point out the still greater danger which threatens the higher or psychic life. I have shown the relation of the mental and spiritual powers and the effects of different foods upon both in a way which I feel sure will convince even the most skeptical of the soundness of my theory.

The time-honored custom of consuming the flesh of animals which has prevailed for so long has flourished unchecked throughout the centuries, in consequence of the popular belief that such a form of diet is both *natural* for man and *necessary*.

In all probability the cruelties which are involved in the flesh traffic, and the wholesale massacre of sentient creatures which flesh-eating necessitates, would have been stopped long ago in Christian lands were it not for the prevalance of this popular delusion. No thoughtful or humane person who is fully acquainted with the nature and extent of these barbarities would attempt to justify them by advancing any other argument than *stern necessity.* As soon, therefore, as this—the only legitimate defence of the custom—is successfully refuted, and Christendom is made to realize that flesh-eating is not only *unnecessary* but also *unnatural* and a violation of Nature's physical laws, the voice of conscience may be trusted to bring about such an awakening of humane sentiment concerning this matter that the era of butchery and bloodshed will ultimately be brought to an end.

Man's physical structure reveals conclusively, and beyond all question, *Nature's intention* concerning his method of living, and *this* revelation is one which is not only irrefutable, tangible and capable of demonstration, but it also demands from us the fullest respect and obedience. *Man is created a "frugivorous" or fruit-eating creature*, and neither his internal organs, his teeth nor his external appearance resemble those of carnivorous animals. To eat flesh, therefore, in any form is to violate a Law of our being which must bring penalty upon us.

It has not been my purpose in this volume to furnish a guide for the cook or housewife, as this has been set forth in the book to which I have previously referred, and which treats wholly on that subject. It can be obtained by sending to the publishers of this book, and should be in every home. I have made a study of the Science of Life, and have published many books dealing with the subjects of Natural Laws and needed reforms in all lines. My work in connection with the National Institute of Science has enabled me to be of personal service to thousands. The books I have written and the course in Personal Magnetism sent out to students have been the means of accomplishing great good.

Those who are desirous of entering upon this natural

way in diet, and are at a loss to know exactly how to begin or have met with difficulties, I would be pleased to have write me, asking any questions that seem necessary. I am also anxious to hear from those who have successfully entered the new life, well knowing that they have found it one of pleasure and happiness.

Thousands of grateful testimonials have been received from those converted after reading the chapter "Flesh Eating a Sin," as published in HOW TO WIN; OR, SURE SECRETS OF SUCCESS. These words of encouragement are a source of pleasure and inspiration to me, and give renewed vigor in this battle with error.

My whole heart is in the work of uplifting humanity, and to that end I have consecrated my life. Trusting that my efforts in the future may be crowned with the success that has been mine in the past,

I am the public's most obedient servant,

L. H. ANDERSON.

BY

PROF. L. H. ANDERSON.

Have you Heard of It? — *A Wonderful Book!*
Have you Seen It? — *A Rare Book!*
Have you Read It? — *A Unique Book!*

JUST THE BOOK to be welcomed by every live, wide-awake person in the land. Brimming over with infectuous Enthusiasm, New Thoughts, New Ideas, to say nothing of old ideas arrayed in a dress so fine you would not recognize them.

This indispensible book will show you in a new light the higher possibilities of life; no one's life a blank; it will stimulate you to new exertions.

Please remember that this remarkable book is not the fine-spun theorizing of a mere dreamer or novelist, but sets forth the practical, every-day common sense ideas of a well-meaning, friendly-disposed business man, who is thoroughly aroused and alive and tingling to his very finger tips with a desire to convey a knowledge of the rarest gifts of earth to his fellow mortals.

Wants. **You have many of them? Know how to gratify them?**

The **Key Note** has been discovered, and it is now given to the world for the **first time.**

YOU MUST HAVE IT

The true "Philosophy of Success"—the art of pleasing, interesting fascinating, science of manners, captivating, making friends of those you meet in society, or every-day social and business life, easily acquired.

Philosophy of one's influence over others; you see it exerted every day by someone in your circle of friends. **You** can do the same. **All** possess the power but do not understand **How to** use it. Develop these latent forces within you, for this faculty of influencing others can be cultivated and developed to an unlimited extent.

YOU, dear reader, possess latent talents which you know not of, and which, if developed, would make you happy and successful far beyond your wildest expectations. It is a man's brain and not his legs that carries him through this world; in brain and not in brawn lies the true power of mankind and his possibilities of development. Knowledge is man's rightful heritage. He needs but reach for it and it is his; he can drink at its everlasting founts and revel in bright realms above those where dwell the human masses. This is no visionary theory but a principle being demonstrated every day. Do no longer hesitate and

◆◆◆◆

remain in darkness regarding these facts, telling you how to develop this faculty to an unlimited extent.

This power has been employed to a limited degree by the learned and brain of all ages—by the African voo-doo; the Chaldean astrologer; the Persian Magi; the Hindoo fakir; the Egyptian priest; the Hebrew prophet, and by the wonder workers of all ages and climes. But it is only very recently that this vast and unlimited power has been fully comprehended and employed with great success in almost every profession, in business or social life. With our aid Personal Magnetism can be developed and made so plain that any intelligent person can learn to employ its invincible power with wonderful success, excelling the marvels of the past as far as day excels the night. Its value counted in dollars and cents can never be estimated. It will be in the near future of greater importance to the human race than all modern inventions which would dazzle into wonders the fantastic dreams of all modern or ancient philosophers. The **secret revealed** so plain that all may understand.

THIS work is the result of years of thought and research. The author has all his life been an absorbed and fascinated student of the mysterious forces which are all powerful in human life; yet are veiled in their operation. Only the earnest seeker after truth who has risen to the higher planes of thought is capable of coming to correct conclusions

regarding the wonderful powers which knowledge and soul development bring within the reach of man, and make subservient to his will.

The inner mysteries of Life are dealt with in this volume in a direct and straightforward way which strips the subject of all illusion and gives the reader the facts, which are now the common property of all mankind. The long night of ignorance and superstition has passed away, and at last the mind of man is free. Reader, do you appreciate the fact that you are living in the latter part of the Nineteenth Century, and that no barrier stands between you and the rich inheritance handed down to you from the ages past?

The truth shall make you free! These words have come down to us from inspired lips, and how little their meaning is known to the masses of men! But every lover of the Occult Sciences has caught a glimpse of the **glorious new era** just now breaking upon humanity. The earnest student of **Personal Magnetism, Hypnotism** and the Occult Sciences never asks the question which is saddening the hearts of agnostics and materialists—**is life worth living?**

You are here informed of secrets which only an honest and upright intention will justify you in using; but we are so sure that in this Science, as in all others, that safety lies in a widespread knowledge and not in ignorance of natural laws, that we fearlessly open to all who desire it the methods and experiments which devoted students of Hypnotism have employed, and the results of their researches.

Professor Anderson has here given his own method of controlling and testing the sensitiveness of subjects to the influence. He reserves only his own instantaneous method, which is his original discovery and is not sent out in book form. To obtain this secret a private course of instructions must be taken, as it is reserved for the students of the College who take the course in Hypnotism.

A special feature of this volume is the beautiful full-page half-tone illustrations; showing the position and expression of subjects while under control; the different attitudes assumed by operator and subjects, etc. The facts and theory of Somnambulism are discussed in a most interesting and lucid manner, and with that straightforward open style that is peculiar to Professor Anderson, and which makes his writings so popular among all classes of people.

We desire, in this connection, to caution the buyer not to place this book in the way of boys or anyone liable to use the information for a bad purpose, as many of the methods are so simple that temptation might overrule better judgment.

Among the many valuable methods disclosed may be mentioned those of Mesmer, Dr. Gregory, Chandos Leigh Hunt, Dr. Darling, Mr. Lewis, Captain Jones, Dr. Keiser, Abbe Farias, Brunos, Deleuze, Billot, Teste, Dupotet, La Fountain, Countess C——, Puysegur, Captain Hudson, Dr. Braid, Prof. Anderson, Dr. Fahenstock, Gruelin, Gassner, Jorden, Bernheim, Moll, Kluge, Hufeland, Prof. Kennedy, Charcot, Choates, Cadwell, Raphael, etc.

Remember, **sixty methods** are given, some of which have sold for $50, but in this rare collection the general average is only 10 cents each, and in addition much other valuable matter along this line is thrown in for good measure.

The information given in this volume relating to animals and how to magnetize them will enable you to train your cat, or dog, or horse, to perform any trick that you wish. It is an open secret that the wonderful power over horses displayed by Rarey and others was simply the power of Magnetism. Price of book only $5.00.

www.ingramcontent.com/pod-product-compliance
Lightning Source LLC
Chambersburg PA
CBHW021514210326
41599CB00012B/1252

* 9 7 8 3 3 3 7 2 0 1 3 4 0 *